Elegant SciPy
The Art of Scientific Python

*Juan Nunez-Iglesias, Stéfan van der Walt,
and Harriet Dashnow*

Beijing · Boston · Farnham · Sebastopol · Tokyo

Elegant SciPy

by Juan Nunez-Iglesias, Stéfan van der Walt, and Harriet Dashnow

Printed in the United States of America.

Published by O'Reilly Media, Inc., 1005 Gravenstein Highway North, Sebastopol, CA 95472.

O'Reilly books may be purchased for educational, business, or sales promotional use. Online editions are also available for most titles (*http://oreilly.com/safari*). For more information, contact our corporate/institutional sales department: 800-998-9938 or corporate@oreilly.com.

Editor: Nan Barber	**Indexer:** Judy McConville
Production Editor: Melanie Yarbrough	**Interior Designer:** David Futato
Copyeditor: Christina Edwards	**Cover Designer:** Karen Montgomery
Proofreader: Rachel Monaghan	**Illustrator:** Rebecca Demarest

August 2017:   First Edition

Revision History for the First Edition

2017-08-10:   First Release

See *http://oreilly.com/catalog/errata.csp?isbn=9781491922873* for release details.

978-1-491-92287-3

[LSI]

Table of Contents

Preface

Unlike the stereotypical wedding dress, it was—to use a technical term—elegant, like a computer algorithm that achieves an impressive outcome with just a few lines of code.
— Graeme Simsion, *The Rosie Effect*

Welcome to *Elegant SciPy*. We're going to spend rather a lot of time focusing on the "SciPy" bit of the title, so let's take a moment to reflect on the "Elegant" bit. There are plenty of manuals, tutorials, and documentation websites out there that describe the SciPy library. *Elegant SciPy* goes further. More than just teaching you how to write code that works, we will inspire you to write code that rocks!

In *The Rosie Effect* (hilarious book; go read its prequel *The Rosie Project* (*https:// en.wikipedia.org/wiki/The_Rosie_Project*) when you're done with *Elegant SciPy*), Graeme Simsion twists the conventions of the word "elegant" around. Most would use it to describe the visual simplicity, style, and grace of, say, the first iPhone. Instead Graeme Simsion's hero, Don Tillman, uses a computer algorithm to *define* elegance. We hope that you will understand exactly what he means after reading this book; that you will read or write a piece of elegant code, and feel calmed in the glow of its beauty and grace. (Note: The authors may be prone to hyperbole.)

A good piece of code just *feels* right. When you look at it, its intent is *clear*, it is often *concise* (but not so concise as to be obscure), and it is *efficient* at executing the task at hand. For the authors, the joy of analyzing elegant code lies in the lessons hidden within, and the way it inspires us to be *creative* in how we approach new coding problems.

Ironically, creativity can also tempt us to show off cleverness at the expense of the reader, and write obtuse code that is hard to understand. PEP8 (the Python style guide) and PEP20 (the Zen of Python) remind us that "code is read much more often than it is written" and therefore "readability counts."

The conciseness of elegant code comes through abstraction and the judicious use of functions, *not* just through packing in a bunch of nested function calls. It may take a minute or two to grok, but it should ultimately provide a crisp, "ah-ha!" moment of understanding. Once you know the various components of the code, its correctness should be obvious. This can be aided by clear variable and function names, and carefully crafted comments that *explain* the code, rather than merely *describe* it.

In the *New York Times*, software engineer J. Bradford Hipps recently argued (*http://nyti.ms/2sEOOwC*) that "to write better code, [one should] read Virginia Woolf":

> As a practice, software development is far more creative than algorithmic.
>
> The developer stands before her source code editor in the same way the author confronts the blank page. [...] They may also share a healthy impatience for the ways things "have always been done" and a generative desire to break conventions. When the module is finished or the pages complete, their quality is judged against many of the same standards: elegance, concision, cohesion; the discovery of symmetries where none were seen to exist. Yes, even beauty.

This is the position we take in this book.

Now that we've dealt with the "elegant" part of the title, let's come back to the "SciPy."

Depending on context, "SciPy" can mean a software library, an ecosystem, or a community. Part of what makes SciPy great is that it has excellent online documentation (*https://docs.scipy.org*) and tutorials (*http://www.scipy-lectures.org*), rendering Just Another Reference book pointless; instead, *Elegant SciPy* wants to present the best code built with SciPy.

The code we have chosen highlights clever, elegant uses of advanced features of NumPy, SciPy, and related libraries. The beginning reader will learn to apply these libraries to real-world problems using beautiful code. And we use real scientific data to motivate our examples.

Like SciPy itself, we wanted *Elegant SciPy* to be driven by the community. We've taken many of our examples from working code found in the wider scientific Python ecosystem, selecting them for their illustration of the principles of elegant code we outlined above.

Who Is This Book For?

Elegant SciPy is intended to inspire you to take your Python to the next level. You will learn SciPy by example, from the very best code.

Before starting, you should at least have seen Python, and know about variables, functions, loops, and maybe a bit of NumPy. You might have even honed your Python skills with advanced material, such as *Fluent Python*. If this doesn't describe you, you should start with some beginner Python tutorials, such as Software Carpentry (*http://software-carpentry.org*), before continuing with this book.

But perhaps you don't know whether the "SciPy stack" is a library or a menu item from the International House of Pancakes, and you aren't sure about best practices. Perhaps you are a scientist who has read some Python tutorials online, and have downloaded some analysis scripts from another lab or a previous member of your own lab, and have fiddled with them. And you might think that you are more or less alone when you learn to code SciPy. You are not.

As we progress, we will teach you how to use the internet as your reference. And we will point you to the mailing lists, repositories, and conferences where you will meet like-minded scientists who are a little further in their journey than you.

This is a book that you will read once, but may return to for inspiration (and maybe to admire some elegant code snippets!).

Why SciPy?

The NumPy and SciPy libraries make up the core of the Scientific Python ecosystem. The SciPy software library implements a set of functions for processing scientific data, such as statistics, signal processing, image processing, and function optimization. SciPy is built on top of NumPy, the Python numerical array computation library. Building on NumPy and SciPy, an entire ecosystem of apps and libraries has grown dramatically over the past few years, spanning a broad spectrum of disciplines that includes astronomy, biology, meteorology and climate science, and materials science, among others.

This growth shows no sign of abating. In 2014, Thomas Robitaille and Chris Beaumont documented (*http://bit.ly/2sF5dRM*) Python's growing use in astronomy. Here's what we found when we updated (*http://bit.ly/2sF5i82*) their plot in the second half of 2016:

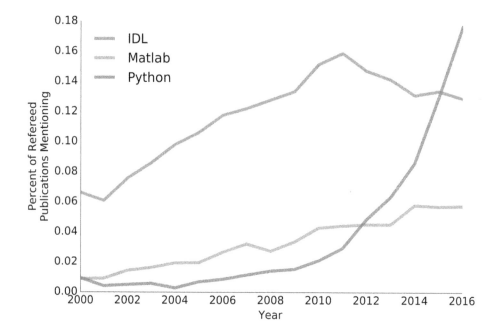

It is clear that SciPy and related libraries will be driving much of scientific data analysis for years to come.

As another example, the Software Carpentry organization (*http://software-carpentry.org*), which teaches computational skills to scientists, most often using Python, currently cannot keep up with demand.

What Is the SciPy Ecosystem?

> *SciPy (pronounced "Sigh Pie") is a Python-based ecosystem of open-source software for mathematics, science, and engineering.*
> —*http://www.scipy.org*

The SciPy ecosystem is a loosely defined collection of Python packages. In *Elegant SciPy*, we will meet many of its main players:

- **NumPy** (*http://www.numpy.org*) is the foundation of scientific computing in Python. It provides efficient numeric arrays and wide support for numerical computation, including linear algebra, random numbers, and Fourier transforms. NumPy's killer feature is its "N-dimensional array," or ndarray. These data structures store numeric values efficiently and define a grid in any number of dimensions (more about this later).

- **SciPy** (*http://www.scipy.org/scipylib/index.html*), the library, is a collection of efficient numerical algorithms for domains such as signal processing, integration, optimization, and statistics. These are wrapped in user-friendly interfaces.
- **Matplotlib** (*http://matplotlib.org*) is a powerful package for plotting in two dimensions (and basic 3D). It draws its name from its Matlab-inspired syntax.
- **IPython** (*https://ipython.org*) is an interactive interface for Python that allows you to quickly interact with your data and test ideas.
- The **Jupyter** (*http://jupyter.org*) notebook runs in your browser and allows the construction of rich documents that combine code, text, mathematical expressions, and interactive widgets.[1] In fact, to produce this book, the text is converted to Jupyter notebooks and executed (that way, we know that all the examples execute correctly). Jupyter started out as an IPython extension, but now supports multiple languages, including Cython, Julia, R, Octave, Bash, Perl, and Ruby.
- **pandas** (*http://pandas.pydata.org*) provides fast, columnar data structures in an easy-to-use package. It is particularly suited to working with labeled datasets such as tables or relational databases, and for managing time series data and sliding windows. pandas also has some handy data analysis tools for data parsing and cleaning, aggregation, and plotting.
- **scikit-learn** (*http://scikit-learn.org*) provides a unified interface to machine learning algorithms.
- **scikit-image** (*http://scikit-image.org*) provides image analysis tools that integrate cleanly with the rest of the SciPy ecosystem.

There are many other Python packages that form part of the SciPy ecosystem, and we will see some of them too. Although this book will focus on NumPy and SciPy, the many surrounding packages are what make Python a powerhouse for scientific computing.

The Great Cataclysm: Python 2 Versus Python 3

In your Python travels, you may have already heard a few rumblings about which version of Python is better. You may have wondered why it's not just the latest version. (Spoiler alert: it is.)

At the end of 2008, the Python core developers released Python 3, a major update to the language with better Unicode (international) text handling, type consistency, and streaming data handling, among other improvements. As Douglas Adams quipped[2] about the creation of the Universe, "this has made a lot of people very angry and been widely regarded as a bad move." That's because Python 2.6 or 2.7 code cannot usually

---

1 Fernando Perez, "'Literate computing' and computational reproducibility: IPython in the age of data-driven journalism" (*http://bit.ly/2sFdfdl*) (blog post), April 19, 2013.

2 Douglas Adams, *The Hitchhiker's Guide to the Galaxy* (London: Pan Books, 1979).

be interpreted by Python 3 without at least some modification (though the changes are typically not too invasive).

There is always a tension between the march of progress and backward compatibility. In this case, the Python core team decided that a clean break was needed to eliminate some inconsistencies, especially in the underlying C API, and moved the language forward into the twenty-first century (Python 1.0 appeared in 1994, more than 20 years ago—a lifetime in the tech world).

Here's one way in which Python has improved in turning 3:

```
print "Hello World!"   # Python 2 print statement
print("Hello World!")  # Python 3 print function
```

Why cause such a fuss just to add some parentheses! Well, true, but what if you want to instead print to a different *stream*, such as *standard error*, the usual place for debugging information?

```
print >>sys.stderr, "fatal error"  # Python 2
print("fatal error", file=sys.stderr)  # Python 3
```

That change certainly seems more worthwhile; what is going on in the Python 2 version anyway? The authors don't rightly know.

Another change is the way Python 3 treats integer division, which is the way most humans treat division. (Note >>> indicates we are typing at the Python interactive shell.)

```
# Python 2
>>> 5 / 2
2
# Python 3
>>> 5 / 2
2.5
```

We were also pretty excited about the new @ *matrix multiplication* operator introduced in Python 3.5 in 2015. Check out Chapters 5 and 6 for some examples of this operator in use!

Possibly the biggest improvement in Python 3 is its support for Unicode, a way of encoding text that allows one to use not just the English alphabet, but any alphabet in the world. Python 2 allowed you to define a Unicode string, like so:

```
beta = u"β"
```

But in Python 3, *everything* is Unicode:

```
β = 0.5
print(2 * β)

1.0
```

The Python core team decided, rightly, that it was worth supporting characters from all languages as first-class citizens in Python code. This is especially true now, when most new coders are from non-English-speaking countries. For the sake of interoperability, we still recommend using English characters in most code, but this capability can come in handy, for example, in math-heavy Jupyter notebooks.

 In the IPython terminal or in the Jupyter notebook, type a LaTeX symbol name followed by the Tab key to have it expanded to Unicode. For example, \beta<TAB> becomes β.

The Python 3 update also breaks a lot of existing 2.x code, and in some cases executes more slowly than before. Despite these frustrations, we encourage all users to upgrade as soon as possible (Python 2.x is now in maintenance only mode until 2020), since most issues have been addressed as the 3.x series has matured. Indeed, we use many new features from Python 3 in this book.

In this book, we use **Python 3.6**.

For more reading, see Ed Schofield's resource, Python-Future, and Nick Coghlan's book-length guide (*http://bit.ly/2sEZoUp*) to the transition.

SciPy Ecosystem and Community

SciPy is a major library with a lot of functionality. Together with NumPy, it is one of Python's killer apps. It has launched a vast number of related libraries that build on this functionality, many of which you'll encounter throughout this book.

The creators of these libraries, and many of their users, gather at many events and conferences around the world. These include the yearly SciPy conference in Austin (USA), EuroSciPy, SciPy India, PyData, and others. We highly recommend attending one of these, and meeting the authors of the best scientific software in the Python world. If you can't get there, or simply want a taste of these conferences, many publish their talks online (*https://www.youtube.com/user/EnthoughtMedia/playlists*).

Free and Open Source Software (FOSS)

The SciPy community embraces open source software development. The source code for nearly all SciPy libraries is freely available to read, edit, and reuse by anyone.

If you want others to use your code, one of the best ways to achieve this is to make it free and open. If you use closed source software, but it doesn't do exactly what you want to achieve, you're out of luck. You can email the developer and ask them to add a new feature (this often doesn't work!), or write new software yourself. If the code is open source, you can easily add or modify its functionality using the skills you learn from this book.

Similarly, if you find a bug in a piece of software, having access to the source code can make things a lot easier for both the user and the developer. Even if you don't quite understand the code, you can usually get a lot further along in diagnosing the problem, and help the developer with fixing it. It is usually a learning experience for everyone!

Open source, open science

In scientific programming, all of the above scenarios are extremely common and important: scientific software often builds on previous work, or modifies it in interesting ways. And, because of the pace of scientific publishing and progress, much code is not thoroughly tested before release, resulting in minor or major bugs.

Another great reason for making code open source is to promote reproducible research. Many of us have had the experience of reading a really cool paper, and then downloading the code to try it out on our own data, only we find that the executable isn't compiled for our system. Or we can't work out how to run it. Or it has bugs, missing features, or produces unexpected results. By making scientific software open source, we not only increase the quality of that software, but we make it possible to see exactly how the science was done. What assumptions were made, and even hard-coded? Open source helps to solve many of these issues. It also enables other scientists to build on the code of their peers, fostering new collaborations and speeding up scientific progress.

Open source licenses

If you want others to use your code, then you *must* license it. If you don't license your code, it is closed by default. Even if you publish your code (e.g., by placing it in a public GitHub repository), without a software license, no one is allowed to use, edit, or redistribute your code.

When choosing among the many license options, you must first decide what you want to allow people to do with your code. Do you want people to be able to sell your code for profit? Or sell software that uses your code? Or do you want to restrict your code to be used only in free software?

There are two broad categories of FOSS licenses:

- Permissive
- Copy-left

A permissive license means you are giving anyone the right to use, edit, and redistribute your code in any way that they like. This includes using your code as part of com-

mercial software. Some popular choices in this category include the MIT and BSD licenses. The SciPy community has adopted the New BSD License (also called "Modified BSD" or "3-clause BSD"). Using such a license means receiving many code contributions from a wide array of people, including many in industry and startups.

Copy-left licenses also allow others to use, edit, and redistribute your code. These licenses, however, also prescribe that derived code must be distributed under a copy-left license. In this way, copy-left licenses restrict what users can do with the code.

The most popular copy-left license is the GNU Public License, or GPL. The main disadvantage to using a copy-left license is that you are often putting your code off-limits to any potential users or contributors from the private sector. And this could include your future self! This can substantially reduce your user base and thus the success of your software. In science, this could mean fewer citations.

For more help choosing a license, see the Choose a License website (*http://chooseali cense.com*). For licensing in a scientific context, we recommend "The Whys and Hows of Licensing Scientific Code," a blog post (*http://bit.ly/2sFj0HS*) by Jake VanderPlas, Director of Research in the Physical Sciences at the University of Washington, and all-around SciPy superstar. In fact, we quote Jake here to drive home the key points of software licensing:

> ...if you only take three pieces of information away from the article, let them be these:
>
> 1. Always license your code. Unlicensed code is closed code, so any open license is better than none (but see #2).
> 2. Always use a GPL-compatible license. GPL-compatible licenses ensure broad compatibility for your code, and include GPL, new BSD, MIT, and others (but see #3).
> 3. Always use a permissive, BSD-style license. A permissive license such as new BSD or MIT is preferable to a copyleft license such as GPL or LGPL.

All the code in this book is available under the 3-Clause BSD license. Where we have sourced code snippets from other people, the code was generally under a permissive open license of some variety (although not necessarily BSD).

For your own code, we recommend that you follow the practices of your community. In scientific Python, this means 3-Clause BSD, while the R language community, for example, has adopted the GPL license.

GitHub: Taking Coding Social

We've talked a little about releasing your source code under an open source license. This will hopefully result in huge numbers of people downloading your code, using it, fixing bugs, and adding new features. Where will you put your code so people can find it? How will those bug fixes and features get back into your code? How will you keep track of all the issues and changes? You can imagine how this could get out of control quite quickly.

Enter GitHub.

GitHub (*https://github.com*) is a website for hosting, sharing, and developing code. It is based on the Git version control software (*http://git-scm.com*). There are some great resources to learn to use GitHub, such as *Introducing GitHub* by Peter Bell and Brent Beer. The vast majority of projects in the SciPy ecosystem are hosted on GitHub, so it is certainly worth learning to use it!

GitHub has had a massive effect on open source contributions. It did this by allowing users to publish code and collaborate for free. Anyone can come along and create a copy (called a *fork*) of the code and edit it to their heart's content. They can eventually contribute those changes back into the original by creating a *pull request*. There are some nice features like managing issues and change requests, as well as the ability to determine who can directly edit your code. You can even keep track of edits, contributors, and other fun stats. There are a whole bunch of other great GitHub features, but we will leave many of them for you to discover and some for you to read in later chapters. In essence, GitHub has democratized software development (Figure P-1), and has substantially reduced the barrier to entry.

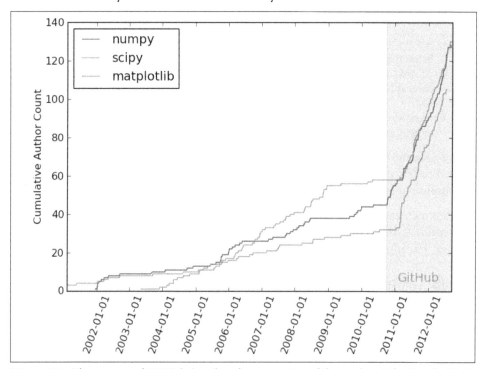

Figure P-1. The impact of GitHub (used with permission of the author, Jake VanderPlas)

Make Your Mark on the SciPy Ecosystem

As you gain more experience with SciPy and start using it for your research, you may find that a particular package is lacking a feature you need, or you think that you can do something more efficiently, or perhaps find a bug. When you reach this point, it's time to start contributing to the SciPy ecosystem.

We strongly encourage you to try doing this. The community lives because people are willing to share their code and improve existing code. And, if we each contribute a little bit, together we build a lot. But, beyond any altruistic reasons for contributing, there are some very practical personal benefits. By engaging with the community you will become a better coder. Any code you contribute will be reviewed by others and you will receive feedback. As a side effect, you will learn how to use Git and GitHub, which are very useful tools for maintaining and sharing your own code. You may even find that interacting with the SciPy community provides you with a broader scientific network and surprising career opportunities.

We want you to think about being more than just a SciPy user. You are joining a community, and your work will make it a better place for all scientific coders.

A Touch of Whimsy with Your Py

In case you were worried that the SciPy community might be an imposing place for the newcomer, remember that it is made of people like you, scientists, usually with a great sense of humor.

In the land of Python, it is inevitable to find some Monty Python references. The package Airspeed Velocity (*http://spacetelescope.github.io/asv/using.html*) measures your software's speed (more on this later), and references the line, "what is the airspeed velocity of an unladen swallow?" from *Monty Python and the Holy Grail*.

Another amusingly titled package is "Sux," which allows you to use Python 2 packages from Python 3. This is a play on "six," which lets you use Python 3 syntax in Python 2, with a New Zealand accent. Sux syntax makes it less frustrating to use Python 2–only packages after you've moved to Python 3:

```
import sux
p = sux.to_use('my_py2_package')
```

In general, Python library names can be a riot, and we hope you'll enjoy your time coming up with some!

Getting Help

Our first step when we get stuck is to Google the task that we are trying to achieve, or the error message that we got. This generally leads us to Stack Overflow (*http://stackoverflow.com/*), an excellent question-and-answer site for programming. If you don't find what you're looking for immediately, try generalizing your search terms to find someone who is having a similar problem.

Sometimes, you might actually be the first person to have this specific question (this is particularly likely when you are using a brand new package), but not all is lost! As mentioned above, the SciPy community is a friendly bunch, and can be found scattered around various parts of the interwebs. Your next point of call is to Google "`<library name> mailing list`," and find an email list to ask for help. Library authors and power users read these regularly, and are very welcoming to newcomers. Note that it is common etiquette to *subscribe* to the list before posting. If you don't, it usually means someone will have to manually check that your email isn't spam before allowing it to be posted to the list. It may seem annoying to join yet another mailing list, but we highly recommend it: it is a great place to learn!

Installing Python

Throughout this book we're going to assume that you have Python 3.6 (or later) and have all the required SciPy packages installed. We list all of the requirements and the versions we used in the *environment.yml* file packaged with the data for this book. The easiest way to get all of these components is to install conda (*http://conda.pydata.org/miniconda.html*), a tool for managing Python environments. You can then pass that *environment.yml* file to conda to install the right versions of everything in one go.

```
conda env create --name elegant-scipy -f path/to/environment.yml
source activate elegant-scipy
```

See the book's GitHub repository (*https://github.com/elegant-scipy/elegant-scipy*) for more details.

Accessing the Book Materials

All of the code and data shown in this book are available on our GitHub repository (*https://github.com/elegant-scipy/elegant-scipy*). In the README file in that repository, you will find instructions to build Jupyter notebooks from the markdown source files, which you can then run interactively using the data included in the repo.

Diving In

We've brought together some of the most elegant code offered up by the SciPy community. Along the way we are going to explore some real-world scientific problems that SciPy solves. This book is also a glimpse into a welcoming, collaborative scientific coding community that wants you to join in.

Welcome to *Elegant SciPy*.

Conventions Used in This Book

The following typographical conventions are used in this book:

Italic
    Indicates new terms, URLs, email addresses, filenames, and file extensions.

`Constant width`
    Used for program listings, as well as within paragraphs to refer to program elements such as variable or function names, databases, data types, environment variables, statements, and keywords.

`Constant width bold`
    Shows commands or other text that should be typed literally by the user.

`Constant width italic`
    Shows text that should be replaced with user-supplied values or by values determined by context.

    This element signifies a tip or suggestion.

    This element signifies a general note.

    This element indicates a warning or caution.

Use of Color

Some of the examples throughout indicate different colors, which is not visible in the print version of this book. Readers of the print book are encouraged to view the source notebooks at *elegant-scipy.org*.

Using Code Examples

Supplemental material (code examples, exercises, etc.) is available for download at *https://github.com/elegant-scipy/elegant-scipy*.

This book is here to help you get your job done. In general, if example code is offered with this book, you may use it in your programs and documentation. You do not need to contact us for permission unless you're reproducing a significant portion of the code. For example, writing a program that uses several chunks of code from this book does not require permission. Selling or distributing a CD-ROM of examples from O'Reilly books does require permission. Answering a question by citing this book and quoting example code does not require permission. Incorporating a significant amount of example code from this book into your product's documentation does require permission.

We appreciate, but do not require, attribution. An attribution usually includes the title, author, publisher, and ISBN. For example: "*Elegant SciPy* by Juan Nunez-Iglesias, Stéfan van der Walt, and Harriet Dashnow (O'Reilly). Copyright 2017 Juan Nunez-Iglesias, Stéfan van der Walt, and Harriet Dashnow, 978-1-491-92287-3."

If you feel your use of code examples falls outside fair use or the permission given above, feel free to contact us at *permissions@oreilly.com*.

O'Reilly Safari

Safari (formerly Safari Books Online) is a membership-based training and reference platform for enterprise, government, educators, and individuals.

Members have access to thousands of books, training videos, Learning Paths, interactive tutorials, and curated playlists from over 250 publishers, including O'Reilly Media, Harvard Business Review, Prentice Hall Professional, Addison-Wesley Professional, Microsoft Press, Sams, Que, Peachpit Press, Adobe, Focal Press, Cisco Press, John Wiley & Sons, Syngress, Morgan Kaufmann, IBM Redbooks, Packt, Adobe Press, FT Press, Apress, Manning, New Riders, McGraw-Hill, Jones & Bartlett, and Course Technology, among others.

For more information, please visit *http://oreilly.com/safari*.

How to Contact Us

Please address comments and questions concerning this book to the publisher:

> O'Reilly Media, Inc.
> 1005 Gravenstein Highway North
> Sebastopol, CA 95472
> 800-998-9938 (in the United States or Canada)
> 707-829-0515 (international or local)
> 707-829-0104 (fax)

To comment or ask technical questions about this book, send email to *bookquestions@oreilly.com*.

For more information about our books, courses, conferences, and news, see our website at http://www.oreilly.com.

Find us on Facebook: *http://facebook.com/oreilly*

Follow us on Twitter: *http://twitter.com/oreillymedia*

Watch us on YouTube: *http://www.youtube.com/oreillymedia*

Acknowledgments

We have to thank the many, many individuals who made essential contributions to this book. It would not have happened without your help.

First and foremost, we wish to thank the many contributors to the NumPy, SciPy, and related libraries. We hope we have done your amazing work justice in this book.

Next, the many contributors to the wider scientific Python ecosystem, including those who provided the foundation for several of our chapters: Vighnesh Birodkar, Matt Rocklin, and Warren Weckesser. We must also thank those whose contributions we were unable to include come press time. Your work inspired us and we hope to include it in future versions of the book. We also thank Nicolas Rougier for his many suggestions that we included as examples and exercises.

Others provided us with data and code that saved us hours of searching and sleuthing. We thank Lav Varshney for the original MATLAB code for spectral graph layout for the worm brain (Chapters 3 and 6), and Stefano Allesina for the St. Marks food web data (Chapter 6).

We are indebted to everyone who made corrections and suggestions while the book was in prerelease, including Bill Katz, Matthias Bussonnier, and Mark Hyun-ki Kim.

We thank our technical reviewers, Thomas Caswell, Nelle Varoquaux, Lav Varshney, and Greg Wilson, who generously took time out of their busy schedules to comb through our final drafts and share their expert advice.

Although we will continue to improve the book based on comments from you, our readers, we owe a great deal to our friends and family who proofread much earlier versions and provided valuable feedback, suggestions, and encouragement. Malcolm Gorman, Alicia Oshack, PW van der Walt, Simon Kocbek, Nelle Varoquaux, and Ariel Rokem: thank you.

And of course, we thank our editors at O'Reilly, Meg Blanchette, Brian MacDonald, and Nan Barber. We are especially grateful to Meg, who first approached us about the book and who offered invaluable early guidance when we had barely a clue what we were doing.

## CHAPTER 1

Elegant NumPy: The Foundation of Scientific Python

[NumPy] is everywhere. It is all around us. Even now, in this very room. You can see it when you look out your window or when you turn on your television. You can feel it when you go to work...when you go to church...when you pay your taxes.
—Morpheus, *The Matrix*

This chapter touches on some statistical functions in SciPy, but more than that, it focuses on exploring the NumPy array, a data structure that underlies almost all numerical scientific computation in Python. We will see how NumPy array operations enable concise and efficient code for manipulating numerical data.

Our use case is using gene expression data from The Cancer Genome Atlas (TCGA) project to predict mortality in skin cancer patients. We will be working toward this goal throughout this chapter and Chapter 2, learning about some key SciPy concepts along the way. Before we can predict mortality, we will need to normalize the expression data using a method called RPKM normalization. This allows the comparison of measurements between different samples and genes. (We will unpack what "gene expression" means in just a moment.)

Let's start with a code snippet to tantalize you and introduce the ideas in this chapter. As we will do in each chapter, we open with a code sample that we believe epitomizes the elegance and power of a particular function from the SciPy ecosystem. In this case, we want to highlight NumPy's vectorization and broadcasting rules, which allow us to manipulate and reason about data arrays very efficiently.

```
def rpkm(counts, lengths):
    """Calculate reads per kilobase transcript per million reads.

    RPKM = (10^9 * C) / (N * L)
```

```
Where:
C = Number of reads mapped to a gene
N = Total mapped reads in the experiment
L = Exon length in base pairs for a gene

Parameters
----------
counts: array, shape (N_genes, N_samples)
    RNAseq (or similar) count data where columns are individual samples
    and rows are genes.
lengths: array, shape (N_genes,)
    Gene lengths in base pairs in the same order
    as the rows in counts.

Returns
-------
normed : array, shape (N_genes, N_samples)
    The RPKM normalized counts matrix.
"""
N = np.sum(counts, axis=0)  # sum each column to get total reads per sample
L = lengths
C = counts

normed = 1e9 * C / (N[np.newaxis, :] * L[:, np.newaxis])

return(normed)
```

This example illustrates some of the ways that NumPy arrays can make your code more elegant:

- Arrays can be 1D, like lists, but they can also be 2D, like matrices, and higher-dimensional still. This allows them to represent many different kinds of numerical data. In our case, we are manipulating a 2D matrix.

- Arrays can be operated on along *axes*. In the first line, we calculate the sum down each column by specifying `axis=0`.

- Arrays allow the expression of many numerical operations at once. For example, toward the end of the function we divide the 2D array of counts (C) by the 1D array of column sums (N). This is broadcasting. More on how this works in just a moment!

Before we delve into the power of NumPy, let's spend some time looking at the biological data that we will be working with.

Introduction to the Data: What Is Gene Expression?

We will work our way through a *gene expression analysis* to demonstrate the power of NumPy and SciPy to solve a real-world biological problem. We will use the pandas library, which builds on NumPy, to read and munge our data files, and then we will manipulate our data efficiently in NumPy arrays.

The so-called central dogma of molecular biology (*https://en.wikipedia.org/wiki/Central_dogma_of_molecular_biology*) states that all the information needed to run a cell (or an organism, for that matter) is stored in a molecule called *deoxyribonucleic acid*, or DNA. This molecule has a repetitive backbone on which lie chemical groups called *bases*, in sequence (Figure 1-1). There are four kinds of bases, abbreviated as A, C, G, and T, comprising an alphabet with which information is stored.

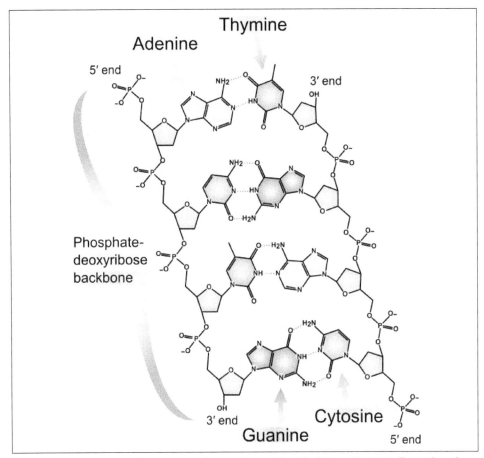

Figure 1-1. The chemical structure of DNA (image by Madeleine Price Ball, used under the terms of the CC0 public domain license)

To access this information, the DNA is *transcribed* into a sister molecule called *messenger ribonucleic acid*, or mRNA. Finally, this mRNA is *translated* into proteins, the workhorses of the cell (Figure 1-2). A section of DNA that encodes the information to make a protein (via mRNA) is called a gene.

The amount of mRNA produced from a given gene is called the *expression* of that gene. Although we would ideally like to measure protein levels, this is a much harder task than measuring mRNA. Fortunately, expression levels of an mRNA and levels of its corresponding protein are usually correlated.[1] Therefore, we usually measure mRNA levels and base our analyses on that. As you will see below, it often doesn't matter, because we are using mRNA levels for their power to predict biological outcomes, rather than to make specific statements about proteins.

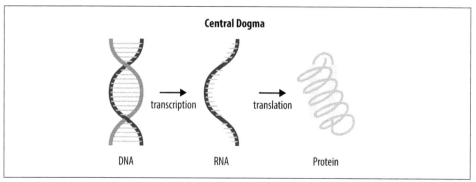

Figure 1-2. Central dogma of molecular biology

It's important to note that the DNA in every cell of your body is identical. Thus, the differences between cells arise from *differential expression* of that DNA into RNA: in different cells, different parts of the DNA are processed into downstream molecules (Figure 1-3). Similarly, as we will see in this chapter and the next, differential expression can distinguish different kinds of cancer.

The state-of-the-art technology to measure mRNA is RNA sequencing (RNAseq). RNA is extracted from a tissue sample (e.g., from a biopsy from a patient), *reverse transcribed* back into DNA (which is more stable), and then read out using chemically modified bases that glow when they are incorporated into the DNA sequence. Currently, high-throughput sequencing machines can only read short fragments (approximately 100 bases is common). These short sequences are called "reads." We measure millions of reads and then based on their sequence we count how many reads came from each gene (Figure 1-4). We'll be starting our analysis directly from this count data.

---

1 Tobias Maier, Marc Güell, and Luis Serrano, "Correlation of mRNA and protein in complex biological samples" (*http://bit.ly/2sFtzLa*), FEBS Letters 583, no. 24 (2009).

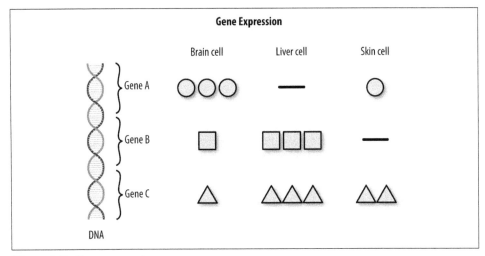

Figure 1-3. Gene expression

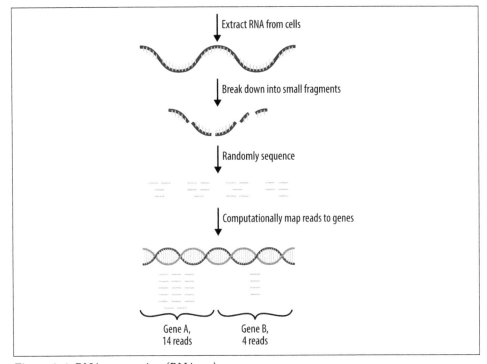

Figure 1-4. RNA sequencing (RNAseq)

Table 1-1 shows a minimal example of a gene expression count data.

Table 1-1. Gene expression count data

	Cell type A	Cell type B
Gene 0	100	200
Gene 1	50	0
Gene 2	350	100

The data is a table of counts, integers representing how many reads were observed for each gene in each cell type. See how the counts for each gene differ between the cell types? We can use this information to learn about the differences between these two types of cells.

One way to represent this data in Python would be as a list of lists:

```
gene0 = [100, 200]
gene1 = [50, 0]
gene2 = [350, 100]
expression_data = [gene0, gene1, gene2]
```

Above, each gene's expression across different cell types is stored in a list of Python integers. Then, we store all of these lists in a list (a *meta-list*, if you will). We can retrieve individual data points using two levels of list indexing:

```
expression_data[2][0]
```

```
350
```

Because of the way the Python interpreter works, this is a very inefficient way to store these data points. First, Python lists are always lists of *objects*, so that the above list gene2 is not a list of integers, but a list of *pointers* to integers, which is unnecessary overhead. Additionally, this means that each of these lists and each of these integers ends up in a completely different, random part of your computer's RAM. However, modern processors actually like to retrieve things from memory in *chunks*, so this spreading of the data throughout the RAM is inefficient.

This is precisely the problem solved by the *NumPy array*.

NumPy N-Dimensional Arrays

One of the key NumPy data types is the N-dimensional array (ndarray, or just array). Ndarrays underpin lots of awesome data manipulation techniques in SciPy. In particular, we're going to explore vectorization and broadcasting, techniques that allow us to write powerful, elegant code to manipulate our data.

First, let's get our heads around the ndarray. These arrays must be homogeneous: all items in an array must be the same type. In our case we will need to store integers. Ndarrays are called N-dimensional because they can have any number of dimensions. A one-dimensional array is roughly equivalent to a Python list:

```
import numpy as np

array1d = np.array([1, 2, 3, 4])
print(array1d)
print(type(array1d))

[1 2 3 4]
<class 'numpy.ndarray'>
```

Arrays have particular attributes and methods you can access by placing a dot after the array name. For example, you can get the array's *shape* with the following:

```
print(array1d.shape)

(4,)
```

Here, it's just a tuple with a single number. You might wonder why you wouldn't just use len, as you would for a list. That will work, but it doesn't extend to *2D* arrays.

This is what we use to represent the data in Table 1-1:

```
array2d = np.array(expression_data)
print(array2d)
print(array2d.shape)
print(type(array2d))

[[100 200]
 [ 50   0]
 [350 100]]
(3, 2)
<class 'numpy.ndarray'>
```

Now you can see that the shape attribute generalizes len to account for the size of multiple dimensions of an array of data.

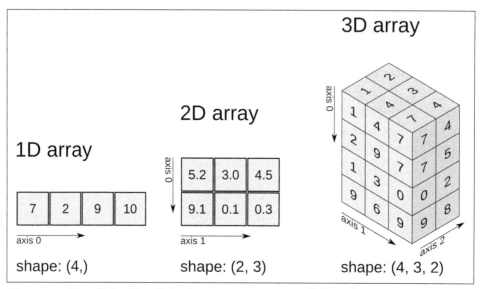

Figure 1-5. Visualizing NumPy's ndarrays in one, two, and three dimensions

Arrays have other attributes, such as ndim, the number of dimensions:

```
print(array2d.ndim)

2
```

You'll become familiar with all of these as you start to use NumPy more for your own data analysis.

NumPy arrays can represent data that has even more dimensions, such as magnetic resonance imaging (MRI) data, which includes measurements within a 3D volume. If we store MRI values over time, we might need a 4D NumPy array.

For now, we'll stick to 2D data. Later chapters will introduce higher-dimensional data and will teach you to write code that works for data of any number of dimensions.

Why Use ndarrays Instead of Python Lists?

Arrays are fast because they enable vectorized operations, written in the low-level language C, that act on the whole array. Say you have a list and you want to multiply every element in the list by five. A standard Python approach would be to write a loop that iterates over the elements of the list and multiply each one by five. However, if your data is instead represented as an array, you can multiply every element in the array by five in a single bound. Behind the scenes, the highly optimized NumPy library is doing the iteration as fast as possible.

```
import numpy as np

# Create an ndarray of integers in the range
# 0 up to (but not including) 1,000,000
array = np.arange(1e6)

# Convert it to a list
list_array = array.tolist()
```

Let's compare how long it takes to multiply all the values in the array by five, using the IPython `timeit` magic function. First, when the data is in a list:

```
%timeit -n10 y = [val * 5 for val in list_array]

10 loops, average of 7: 102 ms +- 8.77 ms per loop (using standard deviation)
```

Now, using NumPy's built-in *vectorized* operations:

```
%timeit -n10 x = array * 5

10 loops, average of 7: 1.28 ms +- 206 µs per loop (using standard deviation)
```

Over 50 times faster, and more concise, too!

Arrays are also size efficient. In Python, each element in a list is an object and is given a healthy memory allocation (or is that unhealthy?). In contrast, in arrays, each element takes up just the necessary amount of memory. For example, an array of 64-bit integers takes up exactly 64 bits per element, plus some very small overhead for array metadata, such as the shape attribute we discussed above. This is generally much less than would be given to objects in a Python list. (If you're interested in digging into how Python memory allocation works, check out Jake VanderPlas's blog post, "Why Python Is Slow: Looking Under the Hood" (*http://bit.ly/2sFDbW8*).)

Plus, when computing with arrays, you can also use *slices* that subset the array *without copying the underlying data*.

```
# Create an ndarray x
x = np.array([1, 2, 3], np.int32)
print(x)

[1 2 3]

# Create a "slice" of x
y = x[:2]
print(y)

[1 2]

# Set the first element of y to be 6
y[0] = 6
print(y)

[6 2]
```

Notice that although we edited y, x has also changed, because y was referencing the same data!

```
# Now the first element in x has changed to 6!
print(x)

[6 2 3]
```

This means you have to be careful with array references. If you want to manipulate the data without touching the original, it's easy to make a copy:

```
y = np.copy(x[:2])
```

Vectorization

Earlier we talked about the speed of operations on arrays. One of the tricks NumPy uses to speed things up is *vectorization*. Vectorization is where you apply a calculation to each element in an array, without having to use a for loop. In addition to speeding things up, this can result in more natural, readable code. Let's look at some examples.

```
x = np.array([1, 2, 3, 4])
print(x * 2)

[2 4 6 8]
```

Here, we have x, an array of 4 values, and we have implicitly multiplied every element in x by 2, a single value.

```
y = np.array([0, 1, 2, 1])
print(x + y)

[1 3 5 5]
```

Now, we have added together each element in x to its corresponding element in y, an array of the same shape.

Both of these operations are simple and, we hope, intuitive examples of vectorization. NumPy also makes them very fast, much faster than iterating over the arrays manually. (Feel free to play with this yourself using the %timeit IPython magic we saw earlier.)

Broadcasting

One of the most powerful and often misunderstood features of ndarrays is broadcasting. Broadcasting is a way of performing implicit operations between two arrays. It allows you to perform operations on arrays of *compatible* shapes, to create arrays bigger than either of the starting ones. For example, we can compute the outer product (*https://en.wikipedia.org/wiki/Outer_product*) of two vectors by reshaping them appropriately:

```
x = np.array([1, 2, 3, 4])
x = np.reshape(x, (len(x), 1))
print(x)

[[1]
 [2]
 [3]
 [4]]

y = np.array([0, 1, 2, 1])
y = np.reshape(y, (1, len(y)))
print(y)

[[0 1 2 1]]
```

Two shapes are compatible when, for each dimension, either is equal to 1 (one) or they match one another.[2]

Let's check the shapes of these two arrays.

```
print(x.shape)
print(y.shape)

(4, 1)
(1, 4)
```

Both arrays have two dimensions and the inner dimensions of both arrays are 1, so the dimensions are compatible!

```
outer = x * y
print(outer)

[[0 1 2 1]
 [0 2 4 2]
 [0 3 6 3]
 [0 4 8 4]]
```

The outer dimensions tell you the size of the resulting array. In our case we expect a (4, 4) array:

```
print(outer.shape)

(4, 4)
```

You can see for yourself that outer[i, j] = x[i] * y[j] for all (i, j).

This was accomplished by NumPy's broadcasting rules (*http://bit.ly/2sFpZ3H*), which implicitly expand dimensions of size 1 in one array to match the corresponding dimension of the other array. Don't worry, we will talk about these rules in more detail later in this chapter.

---

2 We always start by comparing the last dimensions, and work our way forward, ignoring excess dimensions in the case of one array having more than the other (e.g., (3, 5, 1) and (5, 8) would match).

As we will see in the rest of the chapter, as we explore real data, broadcasting is extremely valuable for real-world calculations on arrays of data. It allows us to express complex operations concisely and efficiently.

Exploring a Gene Expression Dataset

The dataset that we'll be using is an RNAseq experiment of skin cancer samples from The Cancer Genome Atlas (TCGA) project (*http://cancergenome.nih.gov*). We've already cleaned and sorted the data for you, so you can just use *data/counts.txt* in the book repository.

In Chapter 2 we will be using this gene expression data to predict mortality in skin cancer patients, reproducing a simplified version of Figures 5A and 5B (*http://bit.ly/2sFCegE*) of a paper (*http://bit.ly/2sFAwfa*) from the TCGA consortium. But first we need to get our heads around the biases in our data, and think about how we could improve it.

Reading in the Data with pandas

We're first going to use pandas to read in the table of counts. pandas is a Python library for data manipulation and analysis, with particular emphasis on tabular and time series data. Here, we will use it to read in tabular data of mixed type. It uses the DataFrame type, which is a flexible tabular format based on the data frame object in R. For example, the data we will read has a column of gene names (strings) and multiple columns of counts (integers), so reading it into a homogeneous array of numbers would be the wrong approach. Although NumPy has some support for mixed data types (called "structured arrays"), it is not primarily designed for this use case, which makes subsequent operations harder than they need to be.

By reading in the data as a pandas data frame we can let pandas do all the parsing, then extract the relevant information and store it in a more efficient data type. Here we are just using pandas briefly to import data. In later chapters we will see a bit more of pandas, but for details, read *Python for Data Analysis* (O'Reilly) by the creator of pandas, Wes McKinney.

```
import numpy as np
import pandas as pd

# Import TCGA melanoma data
filename = 'data/counts.txt'
with open(filename, 'rt') as f:
    data_table = pd.read_csv(f, index_col=0) # Parse file with pandas

print(data_table.iloc[:5, :5])

        00624286-41dd-476f-a63b-d2a5f484bb45  TCGA-FS-A1Z0  TCGA-D9-A3Z1  \
A1BG                                  1272.36        452.96        288.06
```

```
A1CF                                      0.00       0.00       0.00
A2BP1                                     0.00       0.00       0.00
A2LD1                                   164.38     552.43     201.83
A2ML1                                    27.00       0.00       0.00

           02c76d24-f1d2-4029-95b4-8be3bda8fdbe  TCGA-EB-A51B
A1BG                                    400.11     420.46
A1CF                                      1.00       0.00
A2BP1                                     0.00       1.00
A2LD1                                   165.12      95.75
A2ML1                                     0.00       8.00
```

We can see that pandas has kindly pulled out the header row and used it to name the columns. The first column gives the name of each gene, and the remaining columns represent individual samples.

We will also need some corresponding metadata, including the sample information and the gene lengths.

```
# Sample names
samples = list(data_table.columns)
```

We will need some information about the lengths of the genes for our normalization. So that we can take advantage of some fancy pandas indexing, we're going to set the index of the pandas table to be the gene names in the first column.

```
# Import gene lengths
filename = 'data/genes.csv'
with open(filename, 'rt') as f:
    # Parse file with pandas, index by GeneSymbol
    gene_info = pd.read_csv(f, index_col=0)
print(gene_info.iloc[:5, :])
```

```
            GeneID  GeneLength
GeneSymbol
CPA1          1357        1724
GUCY2D        3000        3623
UBC           7316        2687
C11orf95     65998        5581
ANKMY2       57037        2611
```

Let's check how well our gene length data matches up with our count data.

```
print("Genes in data_table: ", data_table.shape[0])
print("Genes in gene_info: ", gene_info.shape[0])

Genes in data_table:  20500
Genes in gene_info:  20503
```

There are more genes in our gene length data than were actually measured in the experiment. Let's filter so we only get the relevant genes, and we want to make sure they are in the same order as in our count data. This is where pandas indexing comes in handy! We can get the intersection of the gene names from our two sources of data and use these to index both datasets, ensuring they have the same genes in the same order.

```
# Subset gene info to match the count data
matched_index = pd.Index.intersection(data_table.index, gene_info.index)
```

Now let's use the intersection of the gene names to index our count data.

```
# 2D ndarray containing expression counts for each gene in each individual
counts = np.asarray(data_table.loc[matched_index], dtype=int)

gene_names = np.array(matched_index)

# Check how many genes and individuals were measured
print(f'{counts.shape[0]} genes measured in {counts.shape[1]} individuals.')

20500 genes measured in 375 individuals.
```

And our gene lengths:

```
# 1D ndarray containing the lengths of each gene
gene_lengths = np.asarray(gene_info.loc[matched_index]['GeneLength'],
                          dtype=int)
```

And let's check the dimensions of our objects:

```
print(counts.shape)
print(gene_lengths.shape)

(20500, 375)
(20500,)
```

As expected, they now match up nicely!

Normalization

Real-world data contains all kinds of measurement artifacts. Before doing any kind of analysis with it, it is important to take a look at it to determine whether some normalization is warranted. For example, measurements with digital thermometers may systematically vary from those taken with mercury thermometers and read by a human. Thus, comparing samples often requires some kind of data wrangling to bring every measurement to a common scale.

In our case, we want to make sure that any differences we uncover correspond to real biological differences, and not to technical artifact. We will consider two levels of normalization often applied jointly to a gene expression dataset: normalization between samples (columns) and normalization between genes (rows).

Between Samples

For example, the number of counts for each individual can vary substantially in RNAseq experiments. Let's take a look at the distribution of expression counts over all the genes. First, we will sum the columns to get the total counts of expression of all genes for each individual, so we can just look at the variation between individuals. To visualize the distribution of total counts, we will use kernel density estimation (KDE), a technique commonly used to smooth out histograms because it gives a clearer picture of the underlying distribution.

Before we start, we have to do some plotting setup (which we will do in every chapter). See "A Quick Note on Plotting" on page 15 for details about each line of the following code.

```
# Make all plots appear inline in the Jupyter notebook from now onwards
%matplotlib inline
# Use our own style file for the plots
import matplotlib.pyplot as plt
plt.style.use('style/elegant.mplstyle')
```

A Quick Note on Plotting

The preceding code does a few neat things to make our plots prettier.

First, `%matplotlib inline` is a Jupyter notebook magic command (*http://bit.ly/2sF9HIb*) that simply makes all plots appear in the notebook rather than pop up a new window. If you are running a Jupyter notebook interactively, you can use `%matplotlib notebook` instead to get an interactive figure, rather than a static image of each plot.

Second, we import `matplotlib.pyplot` and then direct it to use our own plotting style `plt.style.use('style/elegant.mplstyle')`. You will see a block of code like this before the first plot in every chapter.

You may have seen people importing existing styles like this: `plt.style.use('ggplot')`. But we wanted some particular settings, and we wanted all the plots in this book to follow the same style. So we rolled our own Matplotlib style. To see how we did it, take a look at the stylesheet file in the *Elegant SciPy* repository: *style/elegant.mplstyle*. For more information on styles, check out the Matplotlib documentation on stylesheets (*http://bit.ly/2sFz24N*).

Now back to plotting our counts distribution!

```
total_counts = np.sum(counts, axis=0)  # sum columns together
                                        # (axis=1 would sum rows)

from scipy import stats

# Use Gaussian smoothing to estimate the density
density = stats.kde.gaussian_kde(total_counts)

# Make values for which to estimate the density, for plotting
```

```
x = np.arange(min(total_counts), max(total_counts), 10000)

# Make the density plot
fig, ax = plt.subplots()
ax.plot(x, density(x))
ax.set_xlabel("Total counts per individual")
ax.set_ylabel("Density")

plt.show()

print(f'Count statistics:\n  min:  {np.min(total_counts)}'
      f'\n  mean:  {np.mean(total_counts)}'
      f'\n  max:  {np.max(total_counts)}')
```
Count statistics:
  min: 6231205
  mean: 52995255.33866667
  max: 103219262

We can see that there is an order-of-magnitude difference in the total number of counts between the lowest and the highest individual (Figure 1-6). This means that a different number of RNAseq reads were generated for each individual. We say that these individuals have different library sizes.

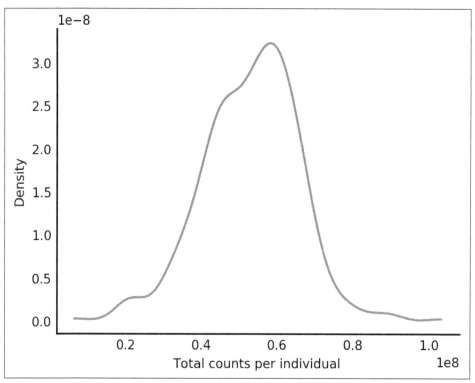

Figure 1-6. Density plot of gene expression counts per individual using KDE smoothing

Normalizing library size between samples

Let's take a closer look at ranges of gene expression for each individual, so when we apply our normalization we can see it in action. We'll subset a random sample of just 70 columns to keep the plotting from getting too messy.

```
# Subset data for plotting
np.random.seed(seed=7) # Set seed so we will get consistent results
# Randomly select 70 samples
samples_index = np.random.choice(range(counts.shape[1]), size=70, replace=False)
counts_subset = counts[:, samples_index]

# Some custom x-axis labelling to make our plots easier to read
def reduce_xaxis_labels(ax, factor):
    """Show only every ith label to prevent crowding on x-axis
        e.g. factor = 2 would plot every second x-axis label,
        starting at the first.

    Parameters
    ----------
    ax : matplotlib plot axis to be adjusted
    factor : int, factor to reduce the number of x-axis labels by
    """
    plt.setp(ax.xaxis.get_ticklabels(), visible=False)
    for label in ax.xaxis.get_ticklabels()[factor-1::factor]:
        label.set_visible(True)

# Bar plot of expression counts by individual
fig, ax = plt.subplots(figsize=(4.8, 2.4))

with plt.style.context('style/thinner.mplstyle'):
    ax.boxplot(counts_subset)
    ax.set_xlabel("Individuals")
    ax.set_ylabel("Gene expression counts")
    reduce_xaxis_labels(ax, 5)
```

There are obviously a lot of outliers at the high expression end of the scale and a lot of variation between individuals, but these are hard to see because everything is clustered around zero (Figure 1-7). So let's do log(n + 1) of our data so it's a bit easier to look at (Figure 1-8). Both the log function and the n + 1 step can be done using broadcasting to simplify our code and speed things up.

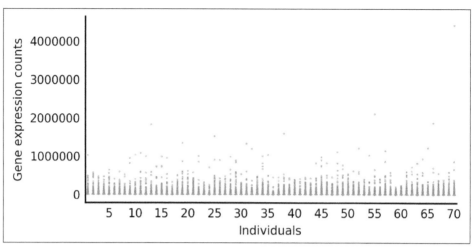

Figure 1-7. Boxplot of gene expression counts per individual

```
# Bar plot of expression counts by individual
fig, ax = plt.subplots(figsize=(4.8, 2.4))

with plt.style.context('style/thinner.mplstyle'):
    ax.boxplot(np.log(counts_subset + 1))
    ax.set_xlabel("Individuals")
    ax.set_ylabel("log gene expression counts")
    reduce_xaxis_labels(ax, 5)
```

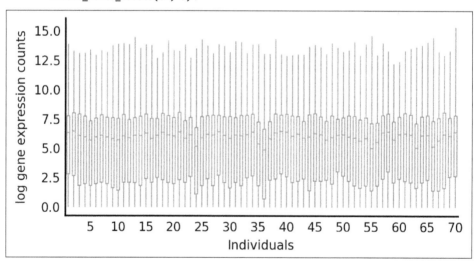

Figure 1-8. Boxplot of gene expression counts per individual (log scale)

Now let's see what happens when we normalize by library size (Figure 1-9).

```
# Normalize by library size
# Divide the expression counts by the total counts for that individual
# Multiply by 1 million to get things back in a similar scale
counts_lib_norm = counts / total_counts * 1000000
# Notice how we just used broadcasting twice there!
counts_subset_lib_norm = counts_lib_norm[:,samples_index]

# Bar plot of expression counts by individual
fig, ax = plt.subplots(figsize=(4.8, 2.4))

with plt.style.context('style/thinner.mplstyle'):
    ax.boxplot(np.log(counts_subset_lib_norm + 1))
    ax.set_xlabel("Individuals")
    ax.set_ylabel("log gene expression counts")
    reduce_xaxis_labels(ax, 5)
```

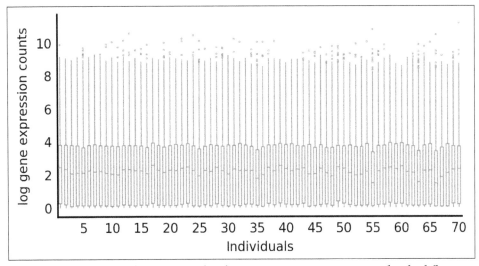

Figure 1-9. Boxplot of library-normalized gene expression counts per individual (log scale)

Much better! Also notice how we used broadcasting twice there. Once to divide all the gene expression counts by the total for that column, and then again to multiply all the values by 1 million.

Finally, let's compare our normalized data to the raw data.

```
import itertools as it
from collections import defaultdict


def class_boxplot(data, classes, colors=None, **kwargs):
    """Make a boxplot with boxes colored according to the class they belong to.

    Parameters
```

```
          ----------
          data : list of array-like of float
              The input data. One boxplot will be generated for each element
              in `data`.
          classes : list of string, same length as `data`
              The class each distribution in `data` belongs to.

          Other parameters
          ----------------
          kwargs : dict
              Keyword arguments to pass on to `plt.boxplot`.
          """
          all_classes = sorted(set(classes))
          colors = plt.rcParams['axes.prop_cycle'].by_key()['color']
          class2color = dict(zip(all_classes, it.cycle(colors)))

          # map classes to data vectors
          # other classes get an empty list at that position for offset
          class2data = defaultdict(list)
          for distrib, cls in zip(data, classes):
              for c in all_classes:
                  class2data[c].append([])
              class2data[cls][-1] = distrib

          # then, do each boxplot in turn with the appropriate color
          fig, ax = plt.subplots()
          lines = []
          for cls in all_classes:
              # set color for all elements of the boxplot
              for key in ['boxprops', 'whiskerprops', 'flierprops']:
                  kwargs.setdefault(key, {}).update(color=class2color[cls])
              # draw the boxplot
              box = ax.boxplot(class2data[cls], **kwargs)
              lines.append(box['whiskers'][0])
          ax.legend(lines, all_classes)
          return ax
```

Now we can plot a colored boxplot according to normalized versus unnormalized samples. We show only three samples from each class for illustration:

```
    log_counts_3 = list(np.log(counts.T[:3] + 1))
    log_ncounts_3 = list(np.log(counts_lib_norm.T[:3] + 1))
    ax = class_boxplot(log_counts_3 + log_ncounts_3,
                       ['raw counts'] * 3 + ['normalized by library size'] * 3,
                       labels=[1, 2, 3, 1, 2, 3])
    ax.set_xlabel('sample number')
    ax.set_ylabel('log gene expression counts');
```

You can see that the normalized distributions are a little more similar when we take library size (the sum of those distributions) into account (Figure 1-10). Now we are comparing like with like between the samples! But what about differences between the genes?

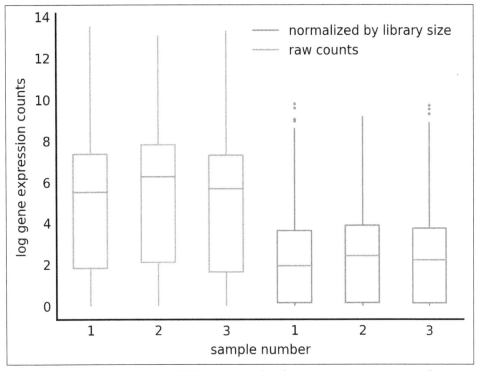

Figure 1-10. Comparing raw and library-normalized gene expression counts in three samples (log scale)

Between Genes

We can also get into some trouble when trying to compare different genes. The number of counts for a gene is related to the gene length. Suppose gene B is twice as long as gene A. Both are expressed at similar levels in the sample (i.e., both produce a similar number of mRNA molecules). Remember that in an RNAseq experiment, we fragment the transcripts and sample reads from that pool of fragments. So if a gene is twice as long, it'll produce twice as many fragments, and we are twice as likely to sample it. Therefore, we would expect gene B to have about twice as many counts as gene A (Figure 1-11). If we want to compare the expression levels of different genes, we will have to do some more normalization.

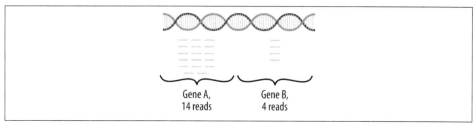

Figure 1-11. Relationship between counts and gene length

Let's see if the relationship between gene length and counts plays out in our dataset. First, we define a utility function for plotting:

```
def binned_boxplot(x, y, *,  # check out this Python 3 exclusive! (*see tip box)
                   xlabel='gene length (log scale)',
                   ylabel='average log counts'):
    """Plot the distribution of `y` dependent on `x` using many boxplots.

    Note: all inputs are expected to be log-scaled.

    Parameters
    ----------
    x: 1D array of float
        Independent variable values.
    y: 1D array of float
        Dependent variable values.
    """
    # Define bins of `x` depending on density of observations
    x_hist, x_bins = np.histogram(x, bins='auto')

    # Use `np.digitize` to number the bins
    # Discard the last bin edge because it breaks the right-open assumption
    # of `digitize`. The max observation correctly goes into the last bin.
    x_bin_idxs = np.digitize(x, x_bins[:-1])

    # Use those indices to create a list of arrays, each containing the `y`
    # values corresponding to `x`s in that bin. This is the input format
    # expected by `plt.boxplot`
    binned_y = [y[x_bin_idxs == i]
                for i in range(np.max(x_bin_idxs))]
    fig, ax = plt.subplots(figsize=(4.8,1))

    # Make the x-axis labels using the bin centers
    x_bin_centers = (x_bins[1:] + x_bins[:-1]) / 2
    x_ticklabels = np.round(np.exp(x_bin_centers)).astype(int)

    # make the boxplot
    ax.boxplot(binned_y, labels=x_ticklabels)

    # show only every 10th label to prevent crowding on x-axis
    reduce_xaxis_labels(ax, 10)
```

```
# Adjust the axis names
ax.set_xlabel(xlabel)
ax.set_ylabel(ylabel);
```

Python 3 Tip: Using * to Create Keyword-Only Arguments

Since version 3.0 Python allows "keyword-only" arguments (*https://www.python.org/dev/peps/pep-3102/*). These are arguments that you have to call using a keyword, rather than relying on position alone. For example, you can call the bin ned_boxplot we just wrote like so:

```
>>> binned_boxplot(x, y, xlabel='my x label', ylabel='my y label')
```

but not like this, which would have been valid Python 2, but raises an error in Python 3:

```
>>> binned_boxplot(x, y, 'my x label', 'my y label')


--------------------------------------------------------------------
TypeError                               Traceback (most recent call last)
<ipython-input-58-7a118d2d5750in <module>()
    1 x_vals = [1, 2, 3, 4, 5]
    2 y_vals = [1, 2, 3, 4, 5]
----3 binned_boxplot(x, y, 'my x label', 'my y label')

TypeError: binned_boxplot() takes 2 positional arguments but 4 were given
```

The idea is to prevent you from accidentally doing something like this:

```
binned_boxplot(x, y, 'my y label')
```

which would give you your y label on the x-axis, a common error for signatures with many optional parameters that don't have obvious ordering.

We now compute the gene lengths and counts:

```
log_counts = np.log(counts_lib_norm + 1)
mean_log_counts = np.mean(log_counts, axis=1)  # across samples
log_gene_lengths = np.log(gene_lengths)

with plt.style.context('style/thinner.mplstyle'):
    binned_boxplot(x=log_gene_lengths, y=mean_log_counts)
```

We can see in the following image that the longer a gene is, the higher its measured counts! As previously explained, this is an artifact of the technique, not a biological signal! How do we account for this?

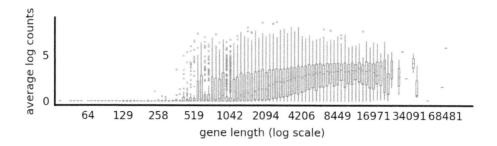

Normalizing Over Samples and Genes: RPKM

One of the simplest normalization methods for RNAseq data is RPKM: reads per kilobase transcript per million reads. RPKM puts together the ideas of normalizing by sample and by gene. When we calculate RPKM, we are normalizing for both the library size (the sum of each column) and the gene length.

To work through how RPKM is derived, let's define the following values:

- C = Number of reads mapped to a gene
- L = Exon length in base-pairs for a gene
- N = Total mapped reads in the experiment

First, let's calculate reads per kilobase.

Reads per base would be:

$$\frac{C}{L}$$

The formula asks for reads per kilobase instead of reads per base. One kilobase = 1,000 bases, so we'll need to divide length (L) by 1,000.

Reads per kilobase would be:

$$\frac{C}{L/1000} = \frac{10^3 C}{L}$$

Next, we need to normalize by library size. If we just divide by the number of mapped reads we get:

$$\frac{10^3 C}{LN}$$

But biologists like thinking in millions of reads so that the numbers don't get too big. Counting per million reads we get:

$$\frac{10^3 C}{L(N/10^6)} = \frac{10^9 C}{LN}$$

In summary, to calculate reads per kilobase transcript per million reads:

$$RPKM = \frac{10^9 C}{LN}$$

Now let's implement RPKM over the entire counts array.

```
# Make our variable names the same as the RPKM formula so we can compare easily
C = counts
N = counts.sum(axis=0)  # sum each column to get total reads per sample
L = gene_lengths  # lengths for each gene, matching rows in `C`
```

First, we multiply by 10^9. Because counts (C) is an ndarray, we can use broadcasting. If we multiple an ndarray by a single value, that value is broadcast over the entire array.

```
# Multiply all counts by 10^9
C_tmp = 10^9 * C
```

Next, we need to divide by the gene length. Broadcasting a single value over a 2D array was pretty clear. We were just multiplying every element in the array by the value. But what happens when we need to divide a 2D array by a 1D array?

Broadcasting rules

Broadcasting allows calculations between ndarrays that have different shapes. Numpy uses broadcasting rules to make these manipulations a little easier. When two arrays have the same number of dimensions, broadcasting can occur if the sizes of each dimension match, or one of them is equal to 1. If arrays have different numbers of dimensions, then (1,) is prepended to the shorter array until the numbers match, and then the standard broadcasting rules apply.

For example, suppose we have two ndarrays, A and B, with shapes (5,2) and (2,). We define the product A * B using broadcasting. B has fewer dimensions than A, so during the calculation, a new dimension is prepended to B with value 1, so B's new shape is (1,2). Finally, where B's shape doesn't match A's, it is *multiplied* by stacking enough versions of B, giving the shape (5,2). This is done "virtually," without using up any additional memory. At this point, the product is just an element-wise multiplication, giving an output array of the same shape as A.

Now let's say we have another array, C, of shape (2.5). To multiply (or add) C to B, we might try to prepend (1,) to the shape of B, but in that case, we still end up with incompatible shapes: (2,5) and (1,2). If we want the arrays to broadcast, we have to *app*end a dimension to B, manually. Then, we end up with (2,5) and (2,1), and broadcasting can proceed.

In NumPy, we can explicitly add a new dimension to B using `np.newaxis`. Let's see this in our normalization by RPKM.

Let's look at the dimensions of our arrays.

```
print('C_tmp.shape', C_tmp.shape)
print('L.shape', L.shape)

C_tmp.shape (20500, 375)
L.shape (20500,)
```

We can see that `C_tmp` has two dimensions, while `L` has one. So during broadcasting, an additional dimension will be prepended to `L`. Then we will have:

```
C_tmp.shape (20500, 375)
L.shape (1, 20500)
```

The dimensions won't match! We want to broadcast `L` over the first dimension of `C_tmp`, so we need to adjust the dimensions of `L` ourselves.

```
L = L[:, np.newaxis] # append a dimension to L, with value 1
print('C_tmp.shape', C_tmp.shape)
print('L.shape', L.shape)

C_tmp.shape (20500, 375)
L.shape (20500, 1)
```

Now that our dimensions match or are equal to one, we can broadcast.

```
# Divide each row by the gene length for that gene (L)
C_tmp = C_tmp / L
```

Finally, we need to normalize by the library size, the total number of counts for that column. Remember that we have already calculated N with:

```
N = counts.sum(axis=0) # sum each column to get total reads per sample

# Check the shapes of C_tmp and N
print('C_tmp.shape', C_tmp.shape)
print('N.shape', N.shape)

C_tmp.shape (20500, 375)
N.shape (375,)
```

Once we trigger broadcasting, an additional dimension will be prepended to N:

```
N.shape (1, 375)
```

The dimensions will match so we don't have to do anything. However, for readability, it can be useful to add the extra dimension to N anyway.

```
# Divide each column by the total counts for that column (N)
N = N[np.newaxis, :]
print('C_tmp.shape', C_tmp.shape)
print('N.shape', N.shape)

C_tmp.shape (20500, 375)
N.shape (1, 375)

# Divide each column by the total counts for that column (N)
rpkm_counts = C_tmp / N
```

Let's put this in a function so we can reuse it.

```
def rpkm(counts, lengths):
    """Calculate reads per kilobase transcript per million reads.

    RPKM = (10^9 * C) / (N * L)

    Where:
    C = Number of reads mapped to a gene
    N = Total mapped reads in the experiment
    L = Exon length in base pairs for a gene

    Parameters
    ----------
    counts: array, shape (N_genes, N_samples)
        RNAseq (or similar) count data where columns are individual samples
        and rows are genes.
    lengths: array, shape (N_genes,)
        Gene lengths in base pairs in the same order
        as the rows in counts.

    Returns
    -------
    normed : array, shape (N_genes, N_samples)
        The RPKM normalized counts matrix.
    """
    N = np.sum(counts, axis=0)  # sum each column to get total reads per sample
    L = lengths
    C = counts

    normed = 1e9 * C / (N[np.newaxis, :] * L[:, np.newaxis])

    return(normed)

counts_rpkm = rpkm(counts, gene_lengths)
```

RPKM between gene normalization

Let's see the RPKM normalization's effect in action. First, as a reminder, here's the distribution of mean log counts as a function of gene length (see Figure 1-12):

```
log_counts = np.log(counts + 1)
mean_log_counts = np.mean(log_counts, axis=1)
log_gene_lengths = np.log(gene_lengths)

with plt.style.context('style/thinner.mplstyle'):
    binned_boxplot(x=log_gene_lengths, y=mean_log_counts)
```

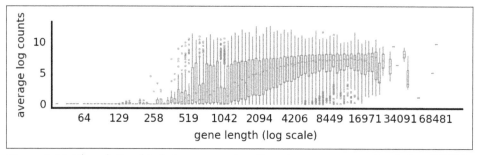

Figure 1-12. The relationship between gene length and average expression before RPKM normalization (log scale)

Now, the same plot with the RPKM-normalized values:

```
log_counts = np.log(counts_rpkm + 1)
mean_log_counts = np.mean(log_counts, axis=1)
log_gene_lengths = np.log(gene_lengths)

with plt.style.context('style/thinner.mplstyle'):
    binned_boxplot(x=log_gene_lengths, y=mean_log_counts)
```

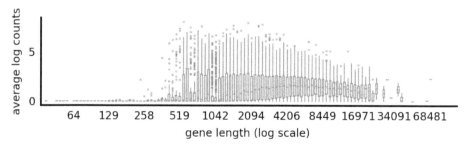

You can see that the mean expression counts have flattened quite a bit, especially for genes larger than about 3,000 base pairs. (Smaller genes still appear to have low expression—these may be too small for the statistical power of the RPKM method.)

RPKM normalization can be useful to compare the expression profile of different genes. We've already seen that longer genes have higher counts, but this doesn't mean their expression level is actually higher. Let's choose a short gene and a long gene and compare their counts before and after RPKM normalization to see what we mean.

```
gene_idxs = np.array([80, 186])
gene1, gene2 = gene_names[gene_idxs]
len1, len2 = gene_lengths[gene_idxs]
gene_labels = [f'{gene1}, {len1}bp', f'{gene2}, {len2}bp']

log_counts = list(np.log(counts[gene_idxs] + 1))
log_ncounts = list(np.log(counts_rpkm[gene_idxs] + 1))

ax = class_boxplot(log_counts,
                   ['raw counts'] * 3,
                   labels=gene_labels)
ax.set_xlabel('Genes')
ax.set_ylabel('log gene expression counts over all samples');
```

If we look at just the raw counts, it looks like the longer gene, TXNDC5, is expressed slightly more than the shorter one, RPL24 (Figure 1-13). But, after RPKM normalization, a different picture emerges:

```
ax = class_boxplot(log_ncounts,
                   ['RPKM normalized'] * 3,
                   labels=gene_labels)
ax.set_xlabel('Genes')
ax.set_ylabel('log RPKM gene expression counts over all samples');
```

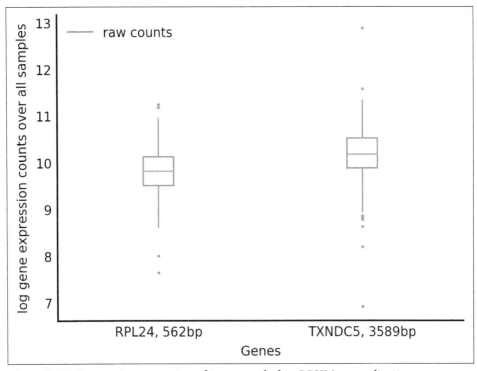

Figure 1-13. Comparing expression of two genes before RPKM normalization

Now it looks like RPL24 is actually expressed at a much higher level than TXNDC5 (see Figure 1-14). This is because RPKM includes normalization for gene length, so we can now directly compare between genes of different lengths.

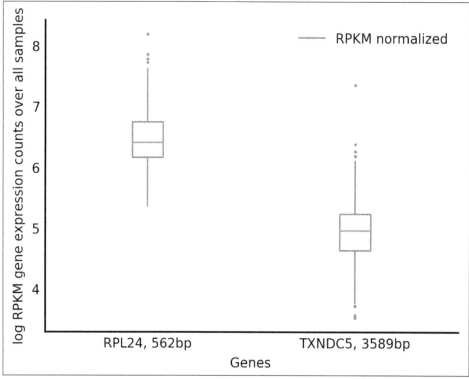

Figure 1-14. Comparing expression of two genes after RPKM normalization

Taking Stock

So far we have done the following:

- Imported data using pandas
- Become familiar with the key NumPy object class—the ndarray
- Used the power of broadcasting to make our calculations more elegant

In Chapter 2 we will continue working with the same dataset, implementing a more sophisticated normalization technique, then use clustering to make some predictions about mortality in skin cancer patients.

Quantile Normalization with NumPy and SciPy

Distress not yourself if you cannot at first understand the deeper mysteries of Spaceland. By degrees they will dawn upon you.

—Edwin A. Abbott, *Flatland: A Romance of Many Dimensions*

In this chapter, we will continue to analyze the gene expression data from Chapter 1, but with a slightly different purpose: we want to use each patient's *gene expression profile* (the full vector of their gene expression measurements) to predict their expected survival. In order to use full profiles, we need a stronger normalization than what Chapter 1's RPKM provides. We will instead perform *quantile normalization* (*https://en.wikipedia.org/wiki/Quantile_normalization*), a technique that ensures measurements fit a specific distribution. This method enforces a strong assumption: if the data are not distributed according to a desired shape, we just make it fit! This might feel a bit like cheating, but it turns out to be simple and useful in many cases where the specific distribution doesn't matter, but the relative changes of values within a population are important. For example, Bolstad and colleagues showed (*http://bit.ly/2tmz3xS*) that it performs admirably in recovering known expression levels in microarray data.

Over the course of the chapter, we will reproduce a simplified version of Figures 5A and 5B (*http://bit.ly/2sFCegE*) from "Genomic Classification of Cutaneous Melanoma," a paper (*http://bit.ly/2sFAwfa*) from The Cancer Genome Atlas (TCGA) project.

Our implementation of quantile normalization uses NumPy and SciPy effectively to produce a function that is fast, efficient, and elegant. Quantile normalization involves three steps:

1. Sort the values along each column
2. Find the average of each resulting row
3. Replace each column quantile with the quantile of the average column

```
import numpy as np
from scipy import stats

def quantile_norm(X):
    """Normalize the columns of X to each have the same distribution.

    Given an expression matrix (microarray data, read counts, etc) of M genes
    by N samples, quantile normalization ensures all samples have the same
    spread of data (by construction).

    The data across each row are averaged to obtain an average column. Each
    column quantile is replaced with the corresponding quantile of the average
    column.

    Parameters
    ----------
    X : 2D array of float, shape (M, N)
        The input data, with M rows (genes/features) and N columns (samples).

    Returns
    -------
    Xn : 2D array of float, shape (M, N)
        The normalized data.
    """
    # compute the quantiles
    quantiles = np.mean(np.sort(X, axis=0), axis=1)

    # compute the column-wise ranks. Each observation is replaced with its
    # rank in that column: the smallest observation is replaced by 1, the
    # second-smallest by 2, ..., and the largest by M, the number of rows.
    ranks = np.apply_along_axis(stats.rankdata, 0, X)

    # convert ranks to integer indices from 0 to M-1
    rank_indices = ranks.astype(int) - 1

    # index the quantiles for each rank with the ranks matrix
    Xn = quantiles[rank_indices]

    return(Xn)
```

Due to the kind of variability of gene expression count data, it is common practice to log-transform the data before quantile-normalizing. Thus, we write an additional helper function to transform to log:

```
def quantile_norm_log(X):
    logX = np.log(X + 1)
    logXn = quantile_norm(logX)
    return logXn
```

Together, these two functions illustrate many of the things that make NumPy powerful (you will remember the first three of these moves from Chapter 1):

- Arrays can be one-dimensional, like lists, but they can also be two-dimensional, like matrices, and higher-dimensional still. This allows them to represent many different kinds of numerical data. In our case, we are representing a 2D matrix.
- Arrays allow the expression of many numerical operations at once. In the first line of `quantile_norm_log`, we add one and take the logarithm for every value in X in a single call. This is called *vectorization*.
- Arrays can be operated on along *axes*. In the first line of `quantile_norm`, we sort the data along each column just by specifying an `axis` parameter to `np.sort`. We then take the mean along each row by specifying a *different* `axis`.
- Arrays underpin the scientific Python ecosystem. The `scipy.stats.rankdata` function operates not on Python lists, but on NumPy arrays. This is true of many scientific libraries in Python.
- Even functions that don't have an `axis=` keyword can be made to operate along axes by NumPy's `apply_along_axis` function.
- Arrays support many kinds of data manipulation through *fancy indexing*: Xn = `quantiles[ranks]`. This is possibly the trickiest part of NumPy, but also among the most useful. We will explore it further in the text that follows.

Getting the Data

As in Chapter 1, we will be working with the TCGA skin cancer RNAseq dataset. Our goal is to predict mortality in skin cancer patients using their RNA expression data. As mentioned earlier, by the end of this chapter we will have reproduced a simplified version of Figures 5A and 5B (*http://bit.ly/2sFCegE*) of a paper (*http://dx.doi.org/10.1016/j.cell.2015.05.044*) from the TCGA consortium.

As in Chapter 1, first we will use pandas to make our job of reading in the data much easier. First we will read in our counts data as a pandas table.

```
import numpy as np
import pandas as pd

# Import TCGA melanoma data
filename = 'data/counts.txt'
data_table = pd.read_csv(filename, index_col=0)  # Parse file with pandas

print(data_table.iloc[:5, :5])

        00624286-41dd-476f-a63b-d2a5f484bb45  TCGA-FS-A1Z0  TCGA-D9-A3Z1  \
A1BG                                  1272.36        452.96        288.06
A1CF                                     0.00          0.00          0.00
A2BP1                                    0.00          0.00          0.00
A2LD1                                  164.38        552.43        201.83
```

A2ML1	27.00	0.00	0.00

	02c76d24-f1d2-4029-95b4-8be3bda8fdbe	TCGA-EB-A51B
A1BG	400.11	420.46
A1CF	1.00	0.00
A2BP1	0.00	1.00
A2LD1	165.12	95.75
A2ML1	0.00	8.00

Looking at the rows and columns of `data_table`, we can see that the columns are the samples, and the rows are the genes. Now let's put our counts in a NumPy array.

```
# 2D ndarray containing expression counts for each gene in each individual
counts = data_table.values
```

Gene Expression Distribution Differences Between Individuals

Now, let's get a feel for our counts data by plotting the distribution of counts for each individual. We will use a Gaussian kernel to smooth out bumps in our data so we can get a better idea of the overall shape.

First, as usual, we set our plotting style:

```
# Make plots appear inline, set custom plotting style
%matplotlib inline
import matplotlib.pyplot as plt
plt.style.use('style/elegant.mplstyle')
```

Next, we write a plotting function that makes use of SciPy's `gaussian_kde` function to plot smooth distributions:

```
from scipy import stats

def plot_col_density(data):
    """For each column, produce a density plot over all rows."""

    # Use Gaussian smoothing to estimate the density
    density_per_col = [stats.gaussian_kde(col) for col in data.T]
    x = np.linspace(np.min(data), np.max(data), 100)

    fig, ax = plt.subplots()
    for density in density_per_col:
        ax.plot(x, density(x))
    ax.set_xlabel('Data values (per column)')
    ax.set_ylabel('Density')
```

Now, we can use that function to plot the distributions of the raw data, before we have done any normalization:

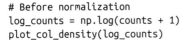

```
# Before normalization
log_counts = np.log(counts + 1)
plot_col_density(log_counts)
```

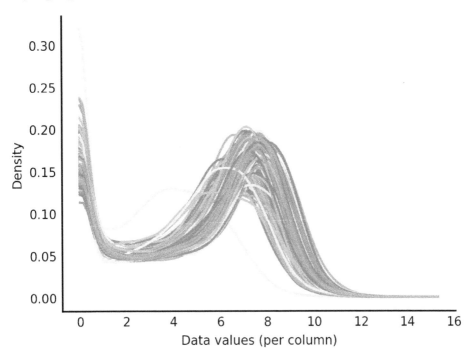

We can see that while the distributions of counts are broadly similar, some individuals have flatter distributions and a few are pushed right over to the left. In fact, realizing that this is a log scale, the location of the peak of the distributions actually varies over an order of magnitude! When doing our analysis of the counts data later in this chapter, we will be assuming that changes in gene expression are due to biological differences between our samples. But a major distribution shift like this suggests that the differences are technical. That is, the changes are likely due to differences in the way we processed each sample, rather than due to biological variation. So we will try to normalize out these global differences between individuals.

To do this normalization, we will perform quantile normalization, as described at the start of the chapter. The idea is that all our samples should have a similar distribution, so any differences in the shape should be due to some technical variation. More formally, given an expression matrix (microarray data, read counts, etc.) of shape (n_genes, n_samples), quantile normalization ensures that all samples (columns) have the same spread of data by construction.

With NumPy and SciPy, this can be done easily and efficiently. To recap, here is our quantile normalization implementation, which we introduced at the beginning of the chapter.

Let's assume we've read in the input matrix as X:

```python
import numpy as np
from scipy import stats

def quantile_norm(X):
    """Normalize the columns of X to each have the same distribution.

    Given an expression matrix (microarray data, read counts, etc.) of M genes
    by N samples, quantile normalization ensures all samples have the same
    spread of data (by construction).

    The data across each row are averaged to obtain an average column. Each
    column quantile is replaced with the corresponding quantile of the average
    column.

    Parameters
    ----------
    X : 2D array of float, shape (M, N)
        The input data, with M rows (genes/features) and N columns (samples).

    Returns
    -------
    Xn : 2D array of float, shape (M, N)
        The normalized data.
    """
    # compute the quantiles
    quantiles = np.mean(np.sort(X, axis=0), axis=1)

    # compute the column-wise ranks. Each observation is replaced with its
    # rank in that column: the smallest observation is replaced by 1, the
    # second-smallest by 2, ..., and the largest by M, the number of rows.
    ranks = np.apply_along_axis(stats.rankdata, 0, X)

    # convert ranks to integer indices from 0 to M-1
    rank_indices = ranks.astype(int) - 1

    # index the quantiles for each rank with the ranks matrix
    Xn = quantiles[rank_indices]

    return(Xn)


def quantile_norm_log(X):
    logX = np.log(X + 1)
    logXn = quantile_norm(logX)
    return logXn
```

Now, let's see what our distributions look like after quantile normalization:

```
# After normalization
log_counts_normalized = quantile_norm_log(counts)

plot_col_density(log_counts_normalized)
```

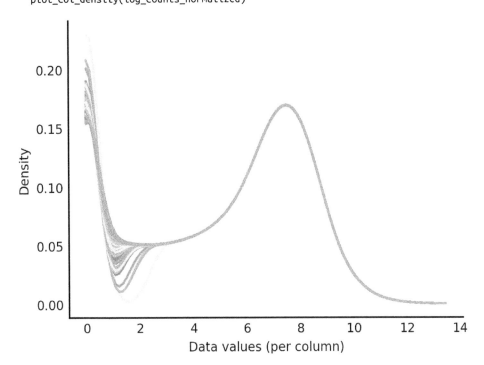

As you might expect, the distributions now look virtually identical! (The different left tails of the distribution have to do with different numbers of ties for low count values —0, 1, 2, ...—in the different columns of the data.)

Now that we have normalized our counts, we can start using our gene expression data to predict patient prognosis.

Biclustering the Counts Data

Clustering the samples tells us which samples have similar gene expression profiles, which may indicate similar characteristics of the samples on other scales. Now that the data are normalized, we can cluster the genes (rows) and samples (columns) of the expression matrix. Clustering the rows tells us which genes' expression values are linked, which is an indication that they work together in the process being studied. *Biclustering* means that we are simultaneously clustering both the rows and columns of our data. By clustering along the rows we find out with genes are working together, and by clustering along the columns we find out which samples are similar.

Because clustering can be an expensive operation, we will limit our analysis to the 1,500 genes that are most variable, since these will account for most of the correlation signal in either dimension.

```
def most_variable_rows(data, *, n=1500):
    """Subset data to the n most variable rows

    In this case, we want the n most variable genes.

    Parameters
    ----------
    data : 2D array of float
        The data to be subset
    n : int, optional
        Number of rows to return.

    Returns
    -------
    variable_data : 2D array of float
        The `n` rows of `data` that exhibit the most variance.
    """
    # compute variance along the columns axis
    rowvar = np.var(data, axis=1)
    # Get sorted indices (ascending order), take the last n
    sort_indices = np.argsort(rowvar)[-n:]
    # use as index for data
    variable_data = data[sort_indices, :]
    return variable_data
```

Next, we need a function to bicluster the data. Normally, you would use a sophisticated clustering algorithm from the scikit-learn (*http://scikit-learn.org*) library for this. In our case, we want to use hierarchical clustering for simplicity and ease of display. The SciPy library happens to have a perfectly good hierarchical clustering module, though it requires a bit of wrangling to get your head around its interface.

As a reminder, hierarchical clustering is a method to group observations using sequential merging of clusters: initially, every observation is its own cluster. Then, the two nearest clusters are repeatedly merged, and then the next two, and so on, until every observation is in a single cluster. This sequence of merges forms a *merge tree*. By cutting the tree at a specific height, we can get a finer or coarser clustering of observations.

The linkage function in scipy.cluster.hierarchy performs a hierarchical clustering of the rows of a matrix, using a particular metric (for example, Euclidean distance, Manhattan distance, or others) and a particular linkage method, the distance between two clusters (for example, the average distance between all the observations in a pair of clusters).

It returns the merge tree as a "linkage matrix," which contains each merge operation along with the distance computed for the merge and the number of observations in the resulting cluster. From the `linkage` documentation:

> A cluster with an index less than n corresponds to one of the n original observations. The distance between clusters `Z[i, 0]` and `Z[i, 1]` is given by `Z[i, 2]`. The fourth value `Z[i, 3]` represents the number of original observations in the newly formed cluster.

Whew! That's a lot of information, but let's dive right in and hopefully you'll get the hang of it rather quickly. First, we define a function, `bicluster`, that clusters both the rows *and* the columns of a matrix:

```
from scipy.cluster.hierarchy import import linkage


def bicluster(data, linkage_method='average', distance_metric='correlation'):
    """Cluster the rows and the columns of a matrix.

    Parameters
    ----------
    data : 2D ndarray
        The input data to bicluster.
    linkage_method : string, optional
        Method to be passed to `linkage`.
    distance_metric : string, optional
        Distance metric to use for clustering. See the documentation
        for ``scipy.spatial.distance.pdist`` for valid metrics.

    Returns
    -------
    y_rows : linkage matrix
        The clustering of the rows of the input data.
    y_cols : linkage matrix
        The clustering of the cols of the input data.
    """
    y_rows = linkage(data, method=linkage_method, metric=distance_metric)
    y_cols = linkage(data.T, method=linkage_method, metric=distance_metric)
    return y_rows, y_cols
```

Simple: we just call `linkage` for the input matrix and also for the *transpose* of that matrix, in which columns become rows and rows become columns.

Visualizing Clusters

Next, we define a function to visualize the output of that clustering. We are going to rearrange the rows and columns of the input data so that similar rows are together and similar columns are together. And we are additionally going to show the merge tree for both rows and columns, displaying which observations belong together for each. The merge trees are presented as dendrograms, with the branch lengths indicating how similar the observations are to each other (shorter = more similar).

As a word of warning, there is a fair bit of hardcoding of parameters going on here. This is difficult to avoid for plotting, where design is often a matter of eyeballing to find the correct proportions.

```python
from scipy.cluster.hierarchy import dendrogram, leaves_list


def clear_spines(axes):
    for loc in ['left', 'right', 'top', 'bottom']:
        axes.spines[loc].set_visible(False)
    axes.set_xticks([])
    axes.set_yticks([])


def plot_bicluster(data, row_linkage, col_linkage,
                   row_nclusters=10, col_nclusters=3):
    """Perform a biclustering, plot a heatmap with dendrograms on each axis.

    Parameters
    ----------
    data : array of float, shape (M, N)
        The input data to bicluster.
    row_linkage : array, shape (M-1, 4)
        The linkage matrix for the rows of `data`.
    col_linkage : array, shape (N-1, 4)
        The linkage matrix for the columns of `data`.
    n_clusters_r, n_clusters_c : int, optional
        Number of clusters for rows and columns.
    """
    fig = plt.figure(figsize=(4.8, 4.8))

    # Compute and plot row-wise dendrogram
    # `add_axes` takes a "rectangle" input to add a subplot to a figure.
    # The figure is considered to have side-length 1 on each side, and its
    # bottom-left corner is at (0, 0).
    # The measurements passed to `add_axes` are the left, bottom, width, and
    # height of the subplot. Thus, to draw the left dendrogram (for the rows),
    # we create a rectangle whose bottom-left corner is at (0.09, 0.1), and
    # measuring 0.2 in width and 0.6 in height.
    ax1 = fig.add_axes([0.09, 0.1, 0.2, 0.6])
    # For a given number of clusters, we can obtain a cut of the linkage
    # tree by looking at the corresponding distance annotation in the linkage
    # matrix.
    threshold_r = (row_linkage[-row_nclusters, 2] +
                   row_linkage[-row_nclusters+1, 2]) / 2
    with plt.rc_context({'lines.linewidth': 0.75}):
        dendrogram(row_linkage, orientation='left',
```

```
                    color_threshold=threshold_r, ax=ax1)
    clear_spines(ax1)

    # Compute and plot column-wise dendrogram
    # See notes above for explanation of parameters to `add_axes`
    ax2 = fig.add_axes([0.3, 0.71, 0.6, 0.2])
    threshold_c = (col_linkage[-col_nclusters, 2] +
                   col_linkage[-col_nclusters+1, 2]) / 2
    with plt.rc_context({'lines.linewidth': 0.75}):
        dendrogram(col_linkage, color_threshold=threshold_c, ax=ax2)
    clear_spines(ax2)

    # Plot data heatmap
    ax = fig.add_axes([0.3, 0.1, 0.6, 0.6])

    # Sort data by the dendrogram leaves
    idx_rows = leaves_list(row_linkage)
    data = data[idx_rows, :]
    idx_cols = leaves_list(col_linkage)
    data = data[:, idx_cols]

    im = ax.imshow(data, aspect='auto', origin='lower', cmap='YlGnBu_r')
    clear_spines(ax)

    # Axis labels
    ax.set_xlabel('Samples')
    ax.set_ylabel('Genes', labelpad=125)

    # Plot legend
    axcolor = fig.add_axes([0.91, 0.1, 0.02, 0.6])
    plt.colorbar(im, cax=axcolor)

    # display the plot
    plt.show()
```

Now we apply these functions to our normalized counts matrix to display row and column clusterings (Figure 2-1).

```
counts_log = np.log(counts + 1)
counts_var = most_variable_rows(counts_log, n=1500)
yr, yc = bicluster(counts_var, linkage_method='ward',
                   distance_metric='euclidean')
with plt.style.context('style/thinner.mplstyle'):
    plot_bicluster(counts_var, yr, yc)
```

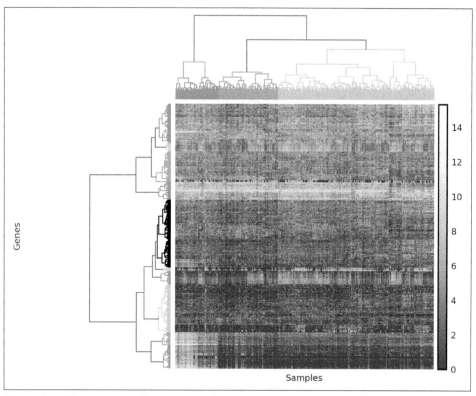

Figure 2-1. This heatmap shows the level of gene expression across all samples and genes. The color indicates the expression level. The rows and columns are grouped by our clusters. We can see our gene clusters along the y-axis and sample clusters across the top of the x-axis.

Predicting Survival

We can see that the sample data naturally falls into at least two clusters, maybe three. Are these clusters meaningful? To answer this, we can access the patient data, available from the data repository (*http://bit.ly/2tiZtR6*) for the paper. After some preprocessing, we get the patients table (*http://bit.ly/2tjp6BD*), which contains survival information for each patient. We can then match these to the counts clusters, and understand whether the patients' gene expression can predict differences in their pathology.

```
patients = pd.read_csv('data/patients.csv', index_col=0)
patients.head()
```

	UV-signature	original-clusters	melanoma-survival-time	melanoma-dead
TCGA-BF-A1PU	UV signature	keratin	NaN	NaN
TCGA-BF-A1PV	UV signature	keratin	13.0	0.0
TCGA-BF-A1PX	UV signature	keratin	NaN	NaN
TCGA-BF-A1PZ	UV signature	keratin	NaN	NaN
TCGA-BF-A1Q0	not UV	immune	17.0	0.0

For each patient (the rows) we have:

UV signature

Ultraviolet light tends to cause specific DNA mutations. By looking for this mutation signature researchers can infer whether UV light likely caused the mutation(s) that led to cancer in these patients.

Original cluster

In the paper, the patients were clustered using gene expression data. These clusters were classified according to the types of genes that typified that cluster. The main clusters were "immune" (n = 168; 51%), "keratin" (n = 102; 31%), and "MITF-low" (n = 59; 18%).

Melanoma survival time

Number of days that the patient survived.

Melanoma dead

One (1) if the patient died of melanoma, zero (0) if they are alive or died of something else.

Now we need to draw *survival curves* for each group of patients defined by the clustering. This is a plot of the fraction of a population that remains alive over a period of time. Note that some data is *right-censored*, which means that in some cases, we don't actually know when the patient died, or the patient might have died of causes unrelated to the melanoma. We count these patients as "alive" for the duration of the survival curve, but more sophisticated analyses might try to estimate their likely time of death.

To obtain a survival curve from survival times, we create a step function that decreases by $1/n$, where n is the number of patients in the group. We then match that function against the noncensored survival times.

```
def survival_distribution_function(lifetimes, right_censored=None):
    """Return the survival distribution function of a set of lifetimes.

    Parameters
    ----------
    lifetimes : array of float or int
```

```
        The observed lifetimes of a population. These must be non-negative.
    right_censored : array of bool, same shape as `lifetimes`
        A value of `True` here indicates that this lifetime was not observed.
        Values of `np.nan` in `lifetimes` are also considered to be
        right-censored.

    Returns
    -------
    sorted_lifetimes : array of float
        The
    sdf : array of float
        Values starting at 1 and progressively decreasing, one level
        for each observation in `lifetimes`.

    Examples
    --------

    In this example, of a population of four, two die at time 1, a
    third dies at time 2, and a final individual dies at an unknown
    time. (Hence, ``np.nan``.)

    >>> lifetimes = np.array([2, 1, 1, np.nan])
    >>> survival_distribution_function(lifetimes)
    (array([ 0.,  1.,  1.,  2.]), array([ 1.  ,  0.75,  0.5 ,  0.25]))
    """
    n_obs = len(lifetimes)
    rc = np.isnan(lifetimes)
    if right_censored is not None:
        rc |= right_censored
    observed = lifetimes[~rc]
    xs = np.concatenate( ([0], np.sort(observed)) )
    ys = np.linspace(1, 0, n_obs + 1)
    ys = ys[:len(xs)]
    return xs, ys
```

Now that we can easily obtain survival curves from the survival data, we can plot
them. We write a function that groups the survival times by cluster identity and plots
each group as a different line:

```
def plot_cluster_survival_curves(clusters, sample_names, patients,
                                 censor=True):
    """Plot the survival data from a set of sample clusters.

    Parameters
    ----------
    clusters : array of int or categorical pd.Series
        The cluster identity of each sample, encoded as a simple int
        or as a pandas categorical variable.
    sample_names : list of string
        The name corresponding to each sample. Must be the same length
        as `clusters`.
    patients : pandas.DataFrame
```

```
    The DataFrame containing survival information for each patient.
    The indices of this DataFrame must correspond to the
    `sample_names`. Samples not represented in this list will be
    ignored.
censor : bool, optional
    If `True`, use `patients['melanoma-dead']` to right-censor the
    survival data.
"""
fig, ax = plt.subplots()
if type(clusters) == np.ndarray:
    cluster_ids = np.unique(clusters)
    cluster_names = ['cluster {}'.format(i) for i in cluster_ids]
elif type(clusters) == pd.Series:
    cluster_ids = clusters.cat.categories
    cluster_names = list(cluster_ids)
n_clusters = len(cluster_ids)
for c in cluster_ids:
    clust_samples = np.flatnonzero(clusters == c)
    # discard patients not present in survival data
    clust_samples = [sample_names[i] for i in clust_samples
                     if sample_names[i] in patients.index]
    patient_cluster = patients.loc[clust_samples]
    survival_times = patient_cluster['melanoma-survival-time'].values
    if censor:
        censored = ~patient_cluster['melanoma-dead'].values.astype(bool)
    else:
        censored = None
    stimes, sfracs = survival_distribution_function(survival_times,
                                                    censored)
    ax.plot(stimes / 365, sfracs)

ax.set_xlabel('survival time (years)')
ax.set_ylabel('fraction alive')
ax.legend(cluster_names)
```

Now we can use the `fcluster` function to obtain cluster identities for the samples (columns of the counts data), and plot each survival curve separately. The `fcluster` function takes a linkage matrix, as returned by `linkage`, and a threshold, and returns cluster identities. It's difficult to know *a priori* what the threshold should be, but we can obtain the appropriate threshold for a fixed number of clusters by checking the distances in the linkage matrix.

```
from scipy.cluster.hierarchy import fcluster
n_clusters = 3
threshold_distance = (yc[-n_clusters, 2] + yc[-n_clusters+1, 2]) / 2
clusters = fcluster(yc, threshold_distance, 'distance')

plot_cluster_survival_curves(clusters, data_table.columns, patients)
```

The clustering of gene expression profiles appears to have identified a higher-risk subtype of melanoma (cluster 2), as shown in Figure 2-2. The TCGA study backs this claim up with a more robust clustering and statistical testing. This is indeed only the latest study to show such a result, with others identifying subtypes of leukemia (blood cancer), gut cancer, and more. Although the above clustering technique is quite fragile, there are other, more robust ways to explore this and similar datasets.[1]

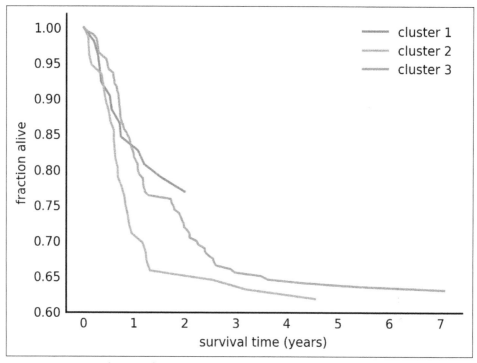

Figure 2-2. Survival curves for patients clustered using gene expression data

Further Work: Using the TCGA's Patient Clusters

Do our clusters do a better job of predicting survival than the original clusters in the paper? What about UV signature? Plot survival curves using the original clusters and UV signature columns of the patient data. How do they compare to our clusters?

Further Work: Reproducing the TCGA's clusters

We leave you the exercise of implementing the approach described in the paper:[2]

---

1 The Cancer Genome Atlas Network, Genomic Classification of Cutaneous Melanoma" (*http://dx.doi.org/10.1016/j.cell.2015.05.044*), *Cell* 161, no. 7 (2015):1681–1696.

2 Ibid.

---

1. Take bootstrap samples (random choice with replacement) of the genes used to cluster the samples.
2. For each sample, produce a hierarchical clustering.
3. In a (n_samples, n_samples)–shaped matrix, store the number of times a sample pair appears together in a bootstrapped clustering.
4. Perform a hierarchical clustering on the resulting matrix.

This identifies groups of samples that frequently occur together in clusterings, regardless of the genes chosen. Thus, these samples can be considered to robustly cluster together.

Hint

Use np.random.choice with replacement=True to create bootstrap samples of row indices.

## CHAPTER 3

Networks of Image Regions with ndimage

Tyger Tyger, burning bright,
In the forests of the night;
What immortal hand or eye,
Could frame thy fearful symmetry?
    —William Blake, *The Tyger*

You probably know that digital images are made up of *pixels*. Generally, you should not think of these as little squares, but as *point samples* of the light signal *measured on a regular grid*.[1]

Further, when processing images, we often deal with objects much larger than individual pixels. In a landscape, the sky, earth, trees, and rocks each span many pixels. A common structure to represent these is the region adjacency graph, or RAG. Its *nodes* hold properties of each region in the image, and its *links* hold the spatial relationships between the regions. Two nodes are linked whenever their corresponding regions touch each other in the input image.

Building such a structure could be a complicated affair, and even more difficult when images are not 2D but 3D and even 4D, as is common in microscopy, materials science, and climatology, among others. But here we will show you how to produce a RAG in a few lines of code using NetworkX (a Python library to analyze graphs and networks), and a filter from SciPy's N-dimensional image processing submodule, ndimage.

```
import networkx as nx
import numpy as np
```

---

1 Alvy Ray Smith, "A Pixel Is Not A Little Square" (*http://alvyray.com/Memos/CG/Microsoft/6_pixel.pdf*), (technical memo) July 17, 1995.

```
from scipy import ndimage as ndi

def add_edge_filter(values, graph):
    center = values[len(values) // 2]
    for neighbor in values:
        if neighbor != center and not graph.has_edge(center, neighbor):
            graph.add_edge(center, neighbor)
    return 0.0


def build_rag(labels, image):
    g = nx.Graph()
    footprint = ndi.generate_binary_structure(labels.ndim, connectivity=1)
    _ = ndi.generic_filter(labels, add_edge_filter, footprint=footprint,
                           mode='nearest', extra_arguments=(g,))
    return g
```

The Origins of Elegant SciPy

(A note from Juan.)

This chapter gets a special mention because it inspired the whole book. Vighnesh Bir-odkar wrote this code snippet as an undergraduate while participating in Google Summer of Code (GSoC) 2014. When I saw this bit of code, it blew me away. For the purposes of this book, it touches on many aspects of scientific Python. By the time you're done with this chapter, you should be able to process arrays of *any* dimension, rather than thinking of them only as 1D lists or 2D tables. More than that, you'll understand the basics of image filtering and network processing.

There are a few things going on here: images being represented as NumPy arrays, *filtering* of these images using `scipy.ndimage`, and building of the image regions into a graph (network) using the NetworkX library. We'll go over these in turn.

Images Are Just NumPy Arrays

In the previous chapter, we saw that NumPy arrays can efficiently represent tabular data, and are a convenient way to perform computations on it. It turns out that arrays are equally adept at representing images.

Here's how to create an image of white noise using just NumPy, and display it with Matplotlib. First, we import the necessary packages, and use the `matplotlib inline` IPython magic to make our images appear below the code:

```
# Make plots appear inline, set custom plotting style
%matplotlib inline
import matplotlib.pyplot as plt
plt.style.use('style/elegant.mplstyle')
```

Then, we "make some noise" and display it as an image:

```
import numpy as np
random_image = np.random.rand(500, 500)
plt.imshow(random_image);
```

This imshow function displays a NumPy array as an image:

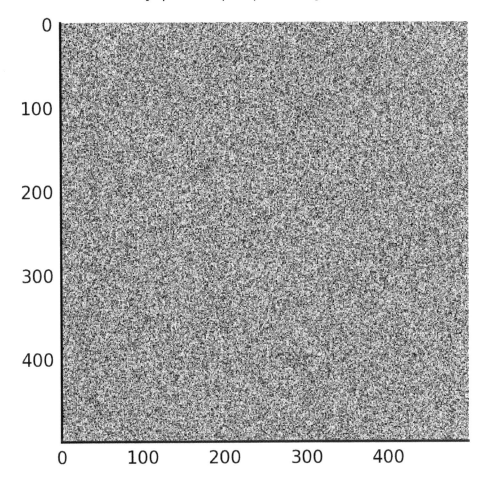

The converse is also true: an image can be considered as a NumPy array. For this example we use the scikit-image library, a collection of image processing tools built on top of NumPy and SciPy.

Here is a PNG image from the scikit-image repository. It is a black and white (sometimes called "grayscale") picture of some ancient Roman coins from Pompeii, obtained from the Brooklyn Museum:

Here is the coin image loaded with scikit-image:

```
from skimage import io
url_coins = ('https://raw.githubusercontent.com/scikit-image/scikit-image/'
             'v0.10.1/skimage/data/coins.png')
coins = io.imread(url_coins)
print("Type:", type(coins), "Shape:", coins.shape, "Data type:", coins.dtype)
plt.imshow(coins);

Type: <class 'numpy.ndarray'> Shape: (303, 384) Data type: uint8
```

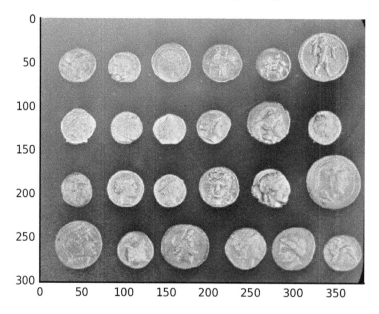

A grayscale image can be represented as a *2D* array, with each array element containing the grayscale intensity at that position. So, *an image is just a NumPy array.*

Color images are *3D* arrays, where the first two dimensions represent the spatial positions of the image, while the final dimension represents color channels, typically the three primary additive colors of red, green, and blue. To show what we can do with these dimensions, let's play with this photo of astronaut Eileen Collins:

```
url_astronaut = ('https://raw.githubusercontent.com/scikit-image/scikit-image/'
                 'master/skimage/data/astronaut.png')
astro = io.imread(url_astronaut)
print("Type:", type(astro), "Shape:", astro.shape, "Data type:", astro.dtype)
plt.imshow(astro);

Type: <class 'numpy.ndarray'> Shape: (512, 512, 3) Data type: uint8
```

This image is *just NumPy arrays*. Adding a green square to the image is easy once you realize this, using simple NumPy slicing:

```
astro_sq = np.copy(astro)
astro_sq[50:100, 50:100] = [0, 255, 0]  # red, green, blue
plt.imshow(astro_sq);
```

You can also use a boolean *mask*, an array of True or False values. We saw these in Chapter 2 as a way to select rows of a table. In this case, we can use an array of the same shape as the image to select pixels:

```
astro_sq = np.copy(astro)
sq_mask = np.zeros(astro.shape[:2], bool)
sq_mask[50:100, 50:100] = True
astro_sq[sq_mask] = [0, 255, 0]
plt.imshow(astro_sq);
```

Exercise: Adding a Grid Overlay

We just saw how to select a square and paint it green. Can you extend that to other shapes and colors? Create a function to draw a blue grid onto a color image, and apply it to the preceding image of Eileen Collins. Your function should take two parameters: the input image and the grid spacing. Use the following template to help you get started:

```
def overlay_grid(image, spacing=128):
    """Return an image with a grid overlay, using the provided spacing.

    Parameters
    ----------
    image : array, shape (M, N, 3)
        The input image.
    spacing : int
        The spacing between the grid lines.
```

```
    Returns
    -------
    image_gridded : array, shape (M, N, 3)
        The original image with a blue grid superimposed.
    """
    image_gridded = image.copy()
    pass  # replace this line with your code...
    return image_gridded
```

```
# plt.imshow(overlay_grid(astro, 128)); # uncomment this line to test your function
```

Check out "Solution: Adding a Grid Overlay" on page 225.

Filters in Signal Processing

Filtering is one of the most fundamental and common operations in image processing. You can filter an image to remove noise, to enhance features, or to detect edges between objects in the image.

To understand filters, it's easiest to start with a 1D signal instead of an image. For example, you might measure the light arriving at your end of a fiber optic cable. If you *sample* the signal every millisecond (ms) for 100ms, you end up with an array of length 100. Suppose that after 30ms the light signal is turned on, and 30ms later, it is switched off. You end up with a signal like this:

```
sig = np.zeros(100, np.float) #
sig[30:60] = 1  # signal = 1 during the period 30-60ms because light is observed
fig, ax = plt.subplots()
ax.plot(sig);
ax.set_ylim(-0.1, 1.1);
```

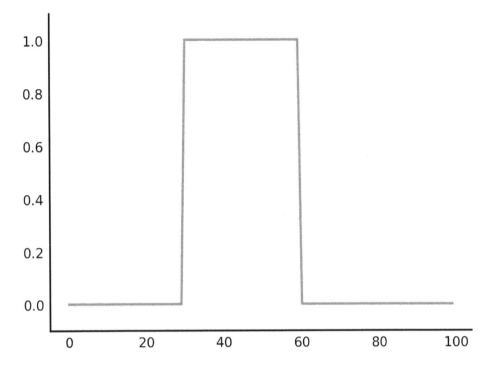

To find *when* the light is turned on, you can *delay* it by 1ms, then *subtract* the original from the delayed signal. This way, when the signal is unchanged from one millisecond to the next, the subtraction will give 0, but when the signal *increases*, you will get a positive signal.

When the signal *decreases*, we will get a negative signal. If we are only interested in pinpointing the time when the light was turned on, we can *clip* the difference signal, so that any negative values are converted to 0:

```
sigdelta = sig[1:]  # sigdelta[0] equals sig[1], and so on
sigdiff = sigdelta - sig[:-1]
sigon = np.clip(sigdiff, 0, np.inf)
fig, ax = plt.subplots()
ax.plot(sigon)
ax.set_ylim(-0.1, 1.1)
print('Signal on at:', 1 + np.flatnonzero(sigon)[0], 'ms')

Signal on at: 30 ms
```

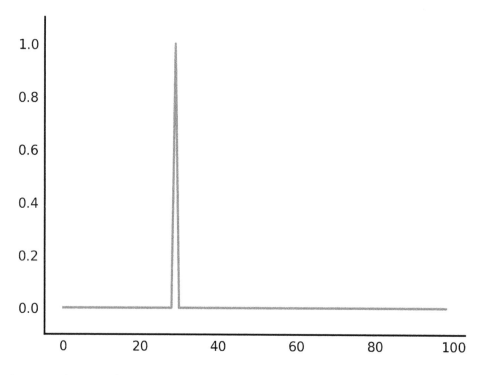

(Here we have used NumPy's `flatnonzero` function to get the first index where the `sigon` array is not equal to 0.)

It turns out that this can be accomplished by a signal processing operation called *convolution*. At every point of the signal, we compute the dot product between the values surrounding that point and a *kernel* or *filter*, which is a predetermined vector of values. Depending on the kernel, then, the convolution shows a different feature of the signal.

Now, think of what happens when the kernel is (1, 0, −1), the difference filter, for a signal `s`. At any position `i`, the convolution result is `1*s[i+1] + 0*s[i] - 1*s[i-1]`, that is, `s[i+1]` - `s[i-1]`. Thus, when the values adjacent to `s[i]` are identical, the convolution gives 0, but when `s[i+1]` > `s[i-1]` (the signal is increasing), it gives a positive value, and, conversely, when `s[i+1]` < `s[i-1]`, it gives a negative value. You can think of this as an estimate of the derivative of the input function.

In general, the formula for convolution is $s'(t) = \Sigma^{t}_{j=t-\tau} s(j) f(t - j)$, where s is the signal, s' is the filtered signal, f is the filter, and τ is the length of the filter.

In SciPy, you can use the `scipy.ndimage.convolve` to work on this:

```
diff = np.array([1, 0, -1])
from scipy import ndimage as ndi
```

```
dsig = ndi.convolve(sig, diff)
plt.plot(dsig);
```

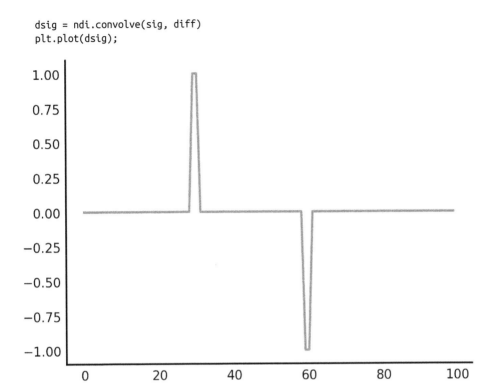

Signals, like those preceding, are usually *noisy* though, not perfect:

```
np.random.seed(0)
sig = sig + np.random.normal(0, 0.3, size=sig.shape)
plt.plot(sig);
```

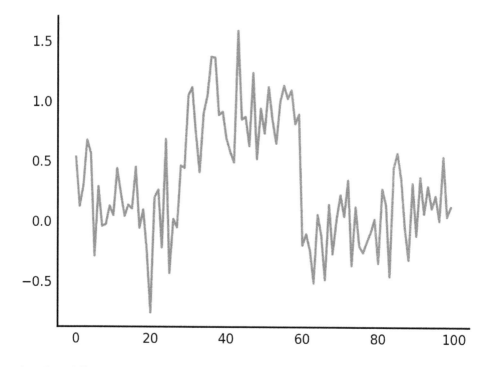

The plain difference filter can amplify that noise:

```
plt.plot(ndi.convolve(sig, diff));
```

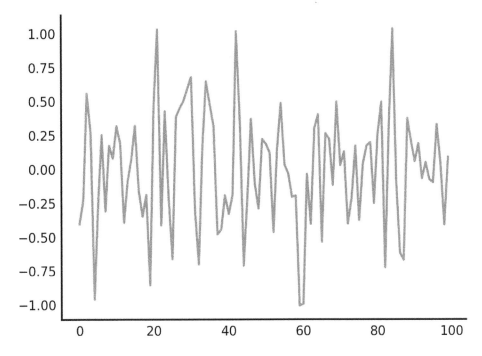

In such cases, you can add smoothing to the filter. The most common form of smoothing is *Gaussian* smoothing, which takes the weighted average of neighboring points in the signal using the Gaussian function (*https://en.wikipedia.org/wiki/Gaussian_function*). We can write a function to make a Gaussian smoothing kernel as follows:

```
def gaussian_kernel(size, sigma):
    """Make a 1D Gaussian kernel of the specified size and standard deviation.

    The size should be an odd number and at least ~6 times greater than sigma
    to ensure sufficient coverage.
    """
    positions = np.arange(size) - size // 2
    kernel_raw = np.exp(-positions**2 / (2 * sigma**2))
    kernel_normalized = kernel_raw / np.sum(kernel_raw)
    return kernel_normalized
```

A really nice feature of convolution is that it's *associative*, meaning if you want to find the derivative of the smoothed signal, you can equivalently convolve the signal with the smoothed difference filter! This can save a lot of computation time, because you can smooth just the filter, which is usually much smaller than the data.

```
smooth_diff = ndi.convolve(gaussian_kernel(25, 3), diff)
plt.plot(smooth_diff);
```

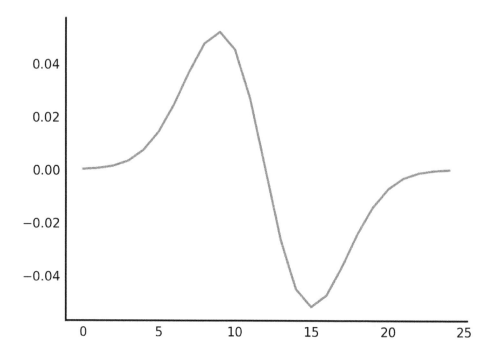

This smoothed difference filter looks for an edge in the central position, but also for that difference to continue. This continuation happens in the case of a true edge, but not in "spurious" edges caused by noise. Check out the result (Figure 3-1):

```
sdsig = ndi.convolve(sig, smooth_diff)
plt.plot(sdsig);
```

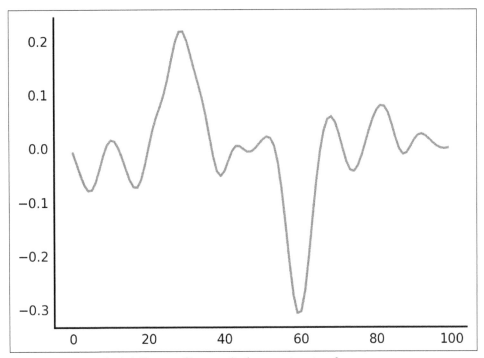

Figure 3-1. Smoothed difference filter applied to a noisy signal

Although it still looks wobbly, the *signal-to-noise ratio* (SNR) is much greater in this version than when we're using the simple difference filter.

Filtering

This operation is called filtering because, in physical electrical circuits, many of these operations are implemented by hardware that allows certain kinds of current through, while blocking others; these hardware components are called filters. For example, a common filter that removes high-frequency voltage fluctuations from a current is called a *low-pass filter*.

Filtering Images (2D Filters)

Now that you've seen filtering in 1D, we hope you'll find it straightforward to extend these concepts to 2D signals, such as images. Here's a 2D difference filter for finding the edges in the coins image:

```
coins = coins.astype(float) / 255  # prevents overflow errors
diff2d = np.array([[0, 1, 0], [1, 0, -1], [0, -1, 0]])
coins_edges = ndi.convolve(coins, diff2d)
io.imshow(coins_edges);
```

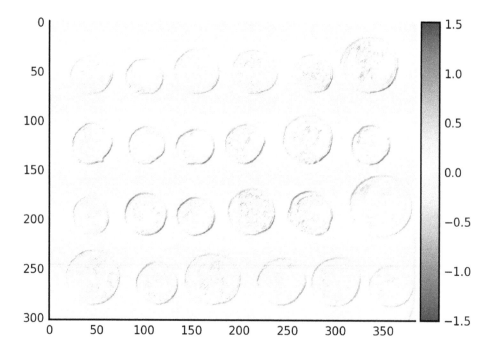

The principle is the same as the 1D filter: at every point in the image, place the filter, compute the dot-product of the filter's values with the image values, and place the result at the same location in the output image. And, as with the 1D difference filter, when the filter is placed on a location with little variation, the dot-product cancels out to zero, whereas, placed on a location where the image brightness is changing, the values multiplied by 1 will be different from those multiplied by –1, and the filtered output will be a positive or negative value (depending on whether the image is brighter toward the bottom right or top left at that point).

Just as with the 1D filter, you can get more sophisticated and smooth out noise right within the filter. The *Sobel* filter is designed to do just that. It comes in horizontal and vertical varieties to find edges with that orientation in the data. Let's start with the horizontal filter first. To find a horizontal edge in a picture, you might try the following filter:

```
# column vector (vertical) to find horizontal edges
hdiff = np.array([[1], [0], [-1]])
```

However, as we saw with 1D filters, this will result in a noisy estimate of the edges in the image. But rather than using Gaussian smoothing, which can cause blurry edges, the Sobel filter uses the property that edges in images tend to be continuous: a picture of the ocean, for example, will contain a horizontal edge along an entire line, not just at specific points of the image. So the Sobel filter smooths the vertical filter horizontally: it looks for a strong edge at the central position that is corroborated by the adjacent positions:

```
hsobel = np.array([[ 1,  2,  1],
                   [ 0,  0,  0],
                   [-1, -2, -1]])
```

The vertical Sobel filter is simply the transpose of the horizontal:

```
vsobel = hsobel.T
```

We can then find the horizontal and vertical edges in the coins image:

```
# Some custom x-axis labeling to make our plots easier to read
def reduce_xaxis_labels(ax, factor):
    """Show only every ith label to prevent crowding on x-axis,
       e.g., factor = 2 would plot every second x-axis label,
       starting at the first.

    Parameters
    ----------
    ax : matplotlib plot axis to be adjusted
    factor : int, factor to reduce the number of x-axis labels by
    """
    plt.setp(ax.xaxis.get_ticklabels(), visible=False)
    for label in ax.xaxis.get_ticklabels()[::factor]:
        label.set_visible(True)


coins_h = ndi.convolve(coins, hsobel)
coins_v = ndi.convolve(coins, vsobel)

fig, axes = plt.subplots(nrows=1, ncols=2)
axes[0].imshow(coins_h, cmap=plt.cm.RdBu)
axes[1].imshow(coins_v, cmap=plt.cm.RdBu)
for ax in axes:
    reduce_xaxis_labels(ax, 2)
```

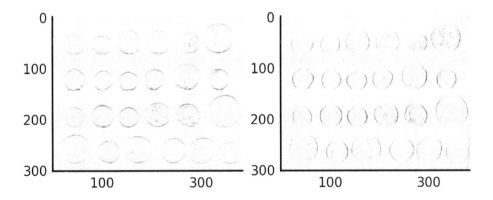

And finally you can argue that, just like the Pythagorean theorem, the edge magnitude in *any* direction is equal to the square root of the sum of squares of the horizontal and vertical components:

```
coins_sobel = np.sqrt(coins_h**2 + coins_v**2)
plt.imshow(coins_sobel, cmap='viridis');
```

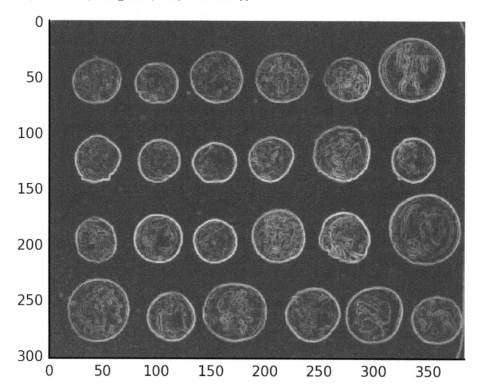

Generic Filters: Arbitrary Functions of Neighborhood Values

In addition to dot-products implemented by ndi.convolve, SciPy lets you define a filter that is an *arbitrary function* of the points in a neighborhood, implemented in ndi.generic_filter. This can let you express arbitrarily complex filters.

For example, suppose an image represents median house values in a county, with a 100m×100m resolution. The local council decides to tax house sales at $10,000 plus 5% of the 90th percentile of house prices in a 1km radius. (So, selling a house in an expensive neighborhood costs more.) With generic_filter, we can produce the map of the tax rate everywhere in the map:

```
from skimage import morphology
def tax(prices):
    return 10000 + 0.05 * np.percentile(prices, 90)
house_price_map = (0.5 + np.random.rand(100, 100)) * 1e6
footprint = morphology.disk(radius=10)
tax_rate_map = ndi.generic_filter(house_price_map, tax, footprint=footprint)
plt.imshow(tax_rate_map)
plt.colorbar();
```

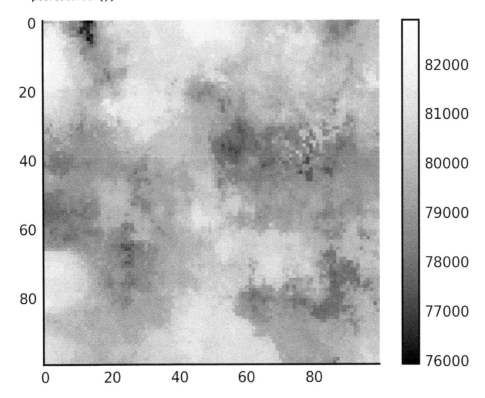

Exercise: Conway's Game of Life

Suggested by Nicolas Rougier

Conway's Game of Life (*https://en.wikipedia.org/wiki/Conway%27s_Game_of_Life*) is a seemingly simple construct in which "cells" on a regular square grid live or die according to the cells in their immediate surroundings. At every timestep, we determine the state of position (i, j) according to its previous state and that of its eight neighbors (above, below, left, right, and diagonals):

- A live cell with only one live neighbor or none dies
- A live cell with two or three live neighbors lives on for another generation
- A live cell with four or more live neighbors dies, as if from overpopulation

- A dead cell with exactly three live neighbors becomes alive, as if by reproduction

Although the rules sound like a contrived math problem, they in fact give rise to incredible patterns, starting with gliders (small patterns of live cells that slowly move in each generation) and glider guns (stationary patterns that sprout off gliders), all the way up to prime number generator machines (e.g., Nathaniel Johnston's "Generating Sequences of Primes in Conway's Game of Life" (*http://bit.ly/2s8UfqF*)), and even simulating Game of Life itself (*https://youtu.be/xP5-iIeKXE8*)!

Can you implement the Game of Life using `ndi.generic_filter`?

Check out "Solution: Conway's Game of Life" on page 226.

Exercise: Sobel Gradient Magnitude

We recently saw how we can combine the output of two different filters, the horizontal Sobel filter and the vertical one. Can you write a function that does this in a single pass using `ndi.generic_filter`?

Check out "Solution: Sobel Gradient Magnitude" on page 227.

Graphs and the NetworkX library

Graphs are a natural representation for an astonishing variety of data. Pages on the web, for example, can comprise nodes, while links between those pages can be, well, links. Or, in biology, so-called *transcription networks* have nodes represent genes and edges connect genes that have a direct influence on each other's expression.

Graphs and Networks

In this context, the term "graph" is synonymous with "network," not with "plot." Mathematicians and computer scientists invented slightly different words to discuss these: graph = network, vertex = node, edge = link = arc. As most people do, we will be using these terms interchangeably.

You might be slightly more familiar with the network terminology: a network consists of *nodes* and *links* between the nodes. Equivalently, a graph consists of *vertices* and *edges* between the vertices. In NetworkX, you have `Graph` objects consisting of `nodes` and `edges` between the nodes, and this is probably the most common usage.

To introduce you to graphs, we will reproduce some results from the paper by Lav Varshney et al., "Structural Properties of the *Caenorhabditis elegans* Neuronal Network" (*http://bit.ly/2s9unuL*).

In our example, we will represent neurons in the nematode worm's nervous system as nodes, and place an edge between two nodes when a neuron makes a synapse with another. (*Synapses* are the chemical connections through which neurons communicate.) The worm is an awesome example of neural connectivity analysis because every worm (of this species) has the same number of neurons (302), and the connections between them are all known. This has resulted in the fantastic Openworm project (*http://www.openworm.org*), which we encourage you to read more about.

You can download the neuronal dataset in Excel format from the WormAtlas database (*http://bit.ly/2s8LmgU*). The `pandas` library allows one to read an Excel table over the web, so we will use it here to read in the data, then feed that into NetworkX.

```
import pandas as pd
connectome_url = 'http://www.wormatlas.org/images/NeuronConnect.xls'
conn = pd.read_excel(connectome_url)
```

conn now contains a pandas `DataFrame`, with rows of the form:

```
[Neuron1, Neuron2, connection type, strength]
```

We are only going to examine the connectome of chemical synapses, so we filter out other synapse types as follows:

```
conn_edges = [(n1, n2, {'weight': s})
              for n1, n2, t, s in conn.itertuples(index=False, name=None)
              if t.startswith('S')]
```

(Look at the WormAtlas page for a description of the different connection types.) We use `weight` in the preceding dictionary because it is a special keyword for edge properties in NetworkX. We then build the graph using NetworkX's `DiGraph` class:

```
import networkx as nx
wormbrain = nx.DiGraph()
wormbrain.add_edges_from(conn_edges)
```

We can now examine some of the properties of this network. One of the first things researchers ask about directed networks is which nodes are the most critical to information flow within it. Nodes with high *betweenness* *centrality* are those that belong to the shortest path between many different pairs of nodes. Think of a rail network: certain stations will connect to many lines, so that you will be forced to change lines there for many different trips. They are the ones with high betweenness centrality.

With NetworkX, we can find similarly important neurons with ease. In the NetworkX API documentation (*http://bit.ly/2tmFvVT*) under "centrality," the docstring for `betweenness_centrality` (*http://bit.ly/2tmhtdC*) specifies a function that takes a graph as input and returns a dictionary mapping node IDs to betweenness centrality values (floating-point values).

```
centrality = nx.betweenness_centrality(wormbrain)
```

Now we can find the neurons with highest centrality using the Python built-in function `sorted`:

```
central = sorted(centrality, key=centrality.get, reverse=True)
print(central[:5])

['AVAR', 'AVAL', 'PVCR', 'PVT', 'PVCL']
```

This returns the neurons AVAR, AVAL, PVCR, PVT, and PVCL, which have been implicated in how the worm responds to prodding: the AVA neurons link the worm's front touch receptors (among others) to neurons responsible for backward motion, while the PVC neurons link the rear touch receptors to forward motion.

Varshney et al. studied the properties of a *strongly connected component* of 237 neurons, out of a total of 279. In graphs, a *connected component* is a set of nodes that are reachable by some path through all the links. The connectome is a *directed* graph, meaning the edges *point* from one node to the other, rather than merely connecting them. In this case, a strongly connected component is one where all nodes are reachable from each other by traversing links *in the correct direction*. So A→B→C is not strongly connected, because there is no way to get to A from B or C. However, A→B→C→A *is* strongly connected.

In a neuronal circuit, you can think of the strongly connected component as the "brain" of the circuit, where the processing happens, while nodes upstream of it are inputs and nodes downstream are outputs.

Cycles in Neuronal Networks

The idea of cyclical neuronal circuits dates back to the 1950s. Here's a lovely paragraph about this idea from an article in *Nautilus*, "The Man Who Tried to Redeem the World with Logic" (*http://bit.ly/2tmmVwZ*), by Amanda Gefter:

If one were to see a lightning bolt flash on the sky, the eyes would send a signal to the brain, shuffling it through a chain of neurons. Starting with any given neuron in the chain, you could retrace the signal's steps and figure out just how long ago the lightning struck. Unless, that is, the chain is a loop. In that case, the information encoding the lightning bolt just spins in circles, endlessly. It bears no connection to the time at which the lightning actually occurred. It becomes, as McCulloch put it, "an idea wrenched out of time." In other words, a memory.

NetworkX makes straightforward work out of getting the largest strongly connected component from our `wormbrain` network:

```
sccs = nx.strongly_connected_component_subgraphs(wormbrain)
giantscc = max(sccs, key=len)
```

```
print(f'The largest strongly connected component has '
      f'{giantscc.number_of_nodes()} nodes, out of '
      f'{wormbrain.number_of_nodes()} total.')
```

```
The largest strongly connected component has 237 nodes, out of 279 total.
```

As noted in the paper, the size of this component is *smaller* than expected by chance, demonstrating that the network is segregated into input, central, and output layers.

Now we reproduce Figure 6B from the paper, the survival function of the in-degree distribution. First, compute the relevant quantities:

```
in_degrees = list(wormbrain.in_degree().values())
in_deg_distrib = np.bincount(in_degrees)
avg_in_degree = np.mean(in_degrees)
cumfreq = np.cumsum(in_deg_distrib) / np.sum(in_deg_distrib)
survival = 1 - cumfreq
```

Then, plot using Matplotlib:

```
fig, ax = plt.subplots()
ax.loglog(np.arange(1, len(survival) + 1), survival)
ax.set_xlabel('in-degree distribution')
ax.set_ylabel('fraction of neurons with higher in-degree distribution')
ax.scatter(avg_in_degree, 0.0022, marker='v')
ax.text(avg_in_degree - 0.5, 0.003, 'mean=%.2f' % avg_in_degree)
ax.set_ylim(0.002, 1.0);
```

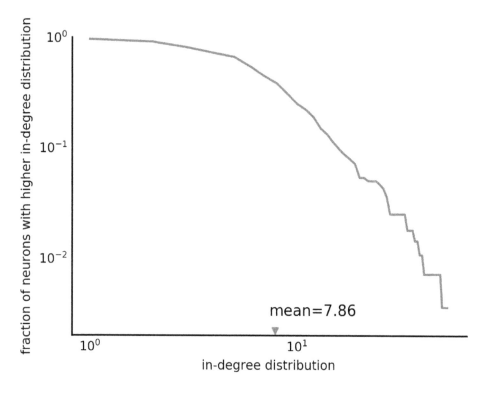

There you have it: a reproduction of a scientific analysis using SciPy. We are missing the line fit...but that's what exercises are for.

Exercise: Curve Fitting with SciPy

This exercise is a bit of a preview for Chapter 7 (optimization): use `scipy.opti mize.curve_fit` to fit the tail of the in-degree survival function to a power law, $f(d) \sim d^{-\gamma}, d > d_0$, for $d_. = 10$, for $d_0 = 10$ (the red line in Figure 6B of the paper), and modify the plot to include that line.

Check out "Solution: Curve Fitting with SciPy" on page 228.

You now should have a fundamental understanding of graphs as a scientific abstraction, and how to easily manipulate and analyze them using Python and NetworkX. Now, we move on to a particular kind of graph used in image processing and computer vision.

Region Adjacency Graphs

A RAG is a representation of an image that is useful for *segmentation*: the division of images into meaningful regions (or *segments*). If you've seen *Terminator 2*, you've seen segmentation (Figure 3-2).

Figure 3-2. Terminator vision

Segmentation is one of those problems that humans do trivially, all the time, without thinking, whereas computers have a hard time of it. To understand this difficulty, look at the following image:

While you see a face, a computer only sees a bunch of numbers:

```
586888888888888899998898988888666532121
668888868889989999998999988888888865421
666655665666899999999999888888888888653
666688999986556889998999888886868665554
668888999988888888899888886656666666543
668888888868686888999888866668888888865
666664433345566888899888666666666668866
668842352214465888899886665644644444666
868644862336646668988665554643212423455
866666583336558888886655655593813663624
888666866686668668888866658588422485434
888888888888686888888866566686666565444
888888886866688888888665566886666868555
888889888888888888888866568886888886666
888889999989999888888888666888888868886
888899988888888888888886566888888888866
888889988888886888888666566868868888888
```

```
68888899988888888886888866568888888888866
68888899999888888886888886556888888888866
68888899986686666888688865656688888888886
88888888886666888888888865655888888888886
68888886665668888889888555555688888888886
86868868658668868688886555555588886866
66688866468866855566655445555656888866
66688654888886868666555554556666666865
88686586888888888866666655556686688665
68888886666888888898888888666666566866665
66888888845686888999888888666655566866655
66688888886245666886666666654431268686655
68688898886689696666665565531366866866655
68888898888668998989998885356888986655
68688889888866899999999866666668986655
68888888888866666888866666666688866655
56888888888686899986868655566688886555
36668888888868888868688666686688866655
26686888888888888888888666688688865654
28688888888888888888666866666868666655
28666688888888888886866866868888866665548
```

Our visual system is so optimized to spot faces that you might see the face even in this blob of numbers! But we hope our point is made. Also, you might want to look at Faces in Things Twitter (*https://twitter.com/facespics*), which demonstrates the face-finding optimization of our visual systems far more humorously.

At any rate, the challenge is to make sense of those numbers, and where the boundaries lie that divide the different parts of the image. A popular approach is to find small regions (called superpixels) that you're *sure* belong in the same segment, and then merge those according to some more sophisticated rule.

As a simple example, suppose you want to segment out the tiger in the following image, from the Berkeley Segmentation Dataset (BSDS).

A clustering algorithm, simple linear iterative clustering (SLIC) (*http://ivrg.epfl.ch/research/superpixels*), can give us a decent starting point. It is available in the scikit-image library.

```
url = ('http://www.eecs.berkeley.edu/Research/Projects/CS/vision/'
       'bsds/BSDS300/html/images/plain/normal/color/108073.jpg')
tiger = io.imread(url)
from skimage import segmentation
seg = segmentation.slic(tiger, n_segments=30, compactness=40.0,
                        enforce_connectivity=True, sigma=3)
```

Scikit-image also has a function to *display* segmentations, which we use to visualize the result of SLIC:

```
from skimage import color
io.imshow(color.label2rgb(seg, tiger));
```

This shows that the body of the tiger has been split in three parts, with the rest of the image in the remaining segments.

A region adjacency graph (RAG) is a graph in which every node represents one of the above regions, and an edge connects two nodes when they touch. For a taste of what it looks like before we build one, we'll use the show_rag function from scikit-image—indeed, the library that contains this chapter's code snippet!

```
from skimage.future import graph

g = graph.rag_mean_color(tiger, seg)
graph.show_rag(seg, g, tiger);
```

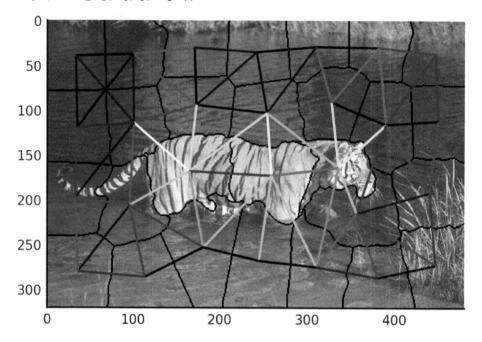

Here, you can see the nodes corresponding to each segment, and the edges between adjacent segments. These are colored with the YlGnBu (yellow-green-blue) colormap from Matplotlib, according to the difference in color between the two nodes.

The preceding figure also shows the magic of thinking of segmentations as graphs: you can see that edges between nodes within the tiger and those outside of it are brighter (higher-valued) than edges within the same object. Thus, if we can cut the graph along those edges, we will get our segmentation. We have chosen an easy example for color-based segmentation, but the same principles hold true for graphs with more complicated pairwise relationships.

Elegant ndimage: How to Build Graphs from Image Regions

All the pieces are in place: you know about NumPy arrays, image filtering, generic filters, graphs, and RAGs. Let's build one to pluck the tiger out of that picture!

The obvious approach is to use two nested for loops to iterate over every pixel of the image, look at the neighboring pixels, and check for different labels:

```
import networkx as nx
def build_rag(labels, image):
    g = nx.Graph()
    nrows, ncols = labels.shape
    for row in range(nrows):
        for col in range(ncols):
            current_label = labels[row, col]
            if not current_label in g:
                g.add_node(current_label)
                g.node[current_label]['total color'] = np.zeros(3, dtype=np.float)
                g.node[current_label]['pixel count'] = 0
            if row < nrows - 1 and labels[row + 1, col] != current_label:
                g.add_edge(current_label, labels[row + 1, col])
            if col < ncols - 1 and labels[row, col + 1] != current_label:
                g.add_edge(current_label, labels[row, col + 1])
            g.node[current_label]['total color'] += image[row, col]
            g.node[current_label]['pixel count'] += 1
    return g
```

Whew! This works, but if you want to segment a 3D image, you'll have to write a different version:

```
import networkx as nx
def build_rag_3d(labels, image):
    g = nx.Graph()
    nplns, nrows, ncols = labels.shape
    for pln in range(nplns):
        for row in range(nrows):
            for col in range(ncols):
                current_label = labels[pln, row, col]
                if not current_label in g:
                    g.add_node(current_label)
                    g.node[current_label]['total color'] = np.zeros(3, dtype=np.float)
                    g.node[current_label]['pixel count'] = 0
                if pln < nplns - 1 and labels[pln + 1, row, col] != current_label:
                    g.add_edge(current_label, labels[pln + 1, row, col])
                if row < nrows - 1 and labels[pln, row + 1, col] != current_label:
                    g.add_edge(current_label, labels[pln, row + 1, col])
                if col < ncols - 1 and labels[pln, row, col + 1] != current_label:
                    g.add_edge(current_label, labels[pln, row, col + 1])
                g.node[current_label]['total color'] += image[pln, row, col]
                g.node[current_label]['pixel count'] += 1
    return g
```

Both of these are pretty ugly and unwieldy, too. And difficult to extend: if we want to count diagonally neighboring pixels as adjacent (i.e., [row, col] is "adjacent to" [row + 1, col + 1]), the code becomes even messier. And if we want to analyze 3D video, we need yet another dimension and another level of nesting. It's a mess!

Enter Vighnesh's insight: SciPy's generic_filter function already does this iteration for us! We used it above to compute an arbitrarily complicated function on the neighborhood of every element of a NumPy array. Only now we don't want a filtered image out of the function: we want a graph. It turns out that `generic_filter` lets you pass additional arguments to the filter function, and we can use that to build the graph:

```python
import networkx as nx
import numpy as np
from scipy import ndimage as nd

def add_edge_filter(values, graph):
    center = values[len(values) // 2]
    for neighbor in values:
        if neighbor != center and not graph.has_edge(center, neighbor):
            graph.add_edge(center, neighbor)
    # float return value is unused but needed by `generic_filter`
    return 0.0

def build_rag(labels, image):
    g = nx.Graph()
    footprint = ndi.generate_binary_structure(labels.ndim, connectivity=1)
    _ = ndi.generic_filter(labels, add_edge_filter, footprint=footprint,
                           mode='nearest', extra_arguments=(g,))
    for n in g:
        g.node[n]['total color'] = np.zeros(3, np.double)
        g.node[n]['pixel count'] = 0
    for index in np.ndindex(labels.shape):
        n = labels[index]
        g.node[n]['total color'] += image[index]
        g.node[n]['pixel count'] += 1
    return g
```

Here are a few reasons this is a brilliant piece of code:

- `ndi.generic_filter` iterates over array elements *with their neighbors*. (Use `numpy.ndindex` to simply iterate over array indices.)
- We return "0.0" from the filter function because `generic_filter` requires the filter function to return a float. However, we ignore the filter output (which is zero everywhere), and use it only for its "side effect" of adding edges to the graph.
- The loops are not nested several levels deep. This makes the code more compact, easier to take in one go.
- The code works identically for 1D, 2D, 3D, or even 8D images!
- If we want to add support for diagonal connectivity, we just need to change the `connectivity` parameter to `ndi.generate_binary_structure`.

Putting It All Together: Mean Color Segmentation

Now, we can use everything we've learned to segment the tiger in the image above:

```python
g = build_rag(seg, tiger)
for n in g:
    node = g.node[n]
```

```
    node['mean'] = node['total color'] / node['pixel count']
for u, v in g.edges_iter():
    d = g.node[u]['mean'] - g.node[v]['mean']
    g[u][v]['weight'] = np.linalg.norm(d)
```

Each edge holds the difference between the average color of each segment. We can
now threshold the graph:

```
def threshold_graph(g, t):
    to_remove = [(u, v) for (u, v, d) in g.edges(data=True)
                 if d['weight'] > t]
    g.remove_edges_from(to_remove)
threshold_graph(g, 80)
```

Finally, we use the NumPy index-with-an-array trick we learned in Chapter 2:

```
map_array = np.zeros(np.max(seg) + 1, int)
for i, segment in enumerate(nx.connected_components(g)):
    for initial in segment:
        map_array[int(initial)] = i
segmented = map_array[seg]
plt.imshow(color.label2rgb(segmented, tiger));
```

Oops! Looks like the cat lost its tail!

Still, we think that's a nice demonstration of the capabilities of RAGs, and the beauty
with which SciPy and NetworkX make it feasible. Many of these functions are available in the scikit-image library. If you are interested in image analysis, look it up!

## CHAPTER 4

Frequency and the Fast Fourier Transform

If you want to find the secrets of the universe, think in terms of energy, frequency and vibration.

    —Nikola Tesla

This chapter was written in collaboration with SW's father, PW van der Walt.

This chapter will depart slightly from the format of the rest of the book. In particular, you may find the *code* in the chapter quite modest. Instead, we want to illustrate an elegant *algorithm*, the Fast Fourier Transform (FFT), that is endlessly useful, is implemented in SciPy, and works, of course, on NumPy arrays.

Introducing Frequency

We'll start by setting up some plotting styles and importing the usual suspects:

```
# Make plots appear inline, set custom plotting style
%matplotlib inline
import matplotlib.pyplot as plt
plt.style.use('style/elegant.mplstyle')

import numpy as np
```

The discrete[1] Fourier transform (DFT) is a mathematical technique used to convert temporal or spatial data into *frequency domain* data. *Frequency* is a familiar concept, due to its colloquial occurrence in the English language: the lowest notes your headphones can rumble out are around 20 Hz, whereas middle C on a piano lies around 261.6 Hz; Hertz, or oscillations per second, in this case literally refers to the number

---

[1] The DFT operates on sampled data, in contrast to the standard Fourier transform, which is defined for continuous functions.

of times per second at which the membrane inside the headphone moves to-and-fro. That, in turn, creates compressed pulses of air which, upon arrival at your eardrum, induces a vibration at the same frequency. So, if you take a simple periodic function, *sin(10 × 2πt)*, you can view it as a wave:

```
f = 10  # Frequency, in cycles per second, or Hertz
f_s = 100  # Sampling rate, or number of measurements per second

t = np.linspace(0, 2, 2 * f_s, endpoint=False)
x = np.sin(f * 2 * np.pi * t)

fig, ax = plt.subplots()
ax.plot(t, x)
ax.set_xlabel('Time [s]')
ax.set_ylabel('Signal amplitude');
```

Or you can equivalently think of it as a repeating signal of *frequency* 10 Hz (it repeats once every 1/10 seconds—a length of time we call its *period*). Although we naturally associate frequency with time, it can equally well be applied to space. For example, a photo of a textile patterns exhibits high *spatial frequency*, whereas the sky or other smooth objects have low spatial frequency.

Let us now examine our sinusoid through application of the DFT:

```
from scipy import fftpack

X = fftpack.fft(x)
freqs = fftpack.fftfreq(len(x)) * f_s

fig, ax = plt.subplots()

ax.stem(freqs, np.abs(X))
ax.set_xlabel('Frequency in Hertz [Hz]')
ax.set_ylabel('Frequency Domain (Spectrum) Magnitude')
ax.set_xlim(-f_s / 2, f_s / 2)
ax.set_ylim(-5, 110)
```

(-5, 110)

We see that the output of the FFT is a 1D array of the same shape as the input, containing complex values. All values are zero, except for two entries. Traditionally, we visualize the magnitude of the result as a *stem plot*, in which the height of each stem corresponds to the underlying value.

(We explain why you see positive and negative frequencies later on in "Discrete Fourier Transforms" on page 95. You may also refer to that section for a more in-depth overview of the underlying mathematics.)

The Fourier transform takes us from the *time* to the *frequency* domain, and this turns out to have a massive number of applications. The *fast Fourier transform* (FFT) is an algorithm for computing the DFT; it achieves its high speed by storing and reusing results of computations as it progresses.

In this chapter, we examine a few applications of the DFT to demonstrate that the FFT can be applied to multidimensional data (not just 1D measurements) to achieve a variety of goals.

Illustration: A Birdsong Spectrogram

Let's start with one of the most common applications, converting a sound signal (consisting of variations of air pressure over time) to a *spectrogram*. You might have seen spectrograms on your music player's equalizer view, or even on an old-school stereo (Figure 4-1).

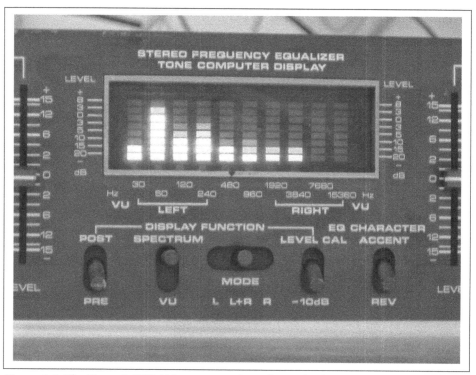

Figure 4-1. The Numark EQ2600 stereo equalizer (http://bit.ly/2s9jRnq) (image used with permission from the author, Sergey Gerasimuk)

Listen to this snippet of the nightingale birdsong (*http://bit.ly/2s9Pq0b*) (released under CC BY 4.0):

```
from IPython.display import Audio
Audio('data/nightingale.wav')
```

If you are reading the paper version of this book, you'll have to use your imagination! It goes something like this: chee-chee-woorrrr-hee-hee cheet-wheet-hoorrr-chirrr-whi-wheo-wheo-wheo-wheo-wheo-wheo.

Since we realize that not everyone is fluent in bird-speak, perhaps it's best if we visualize the measurements—better known as "the signal"—instead.

We load the audio file, which gives us the sampling rate (number of measurements per second) as well as audio data as an (N, 2) array—two columns because this is a stereo recording.

```
from scipy.io import wavfile

rate, audio = wavfile.read('data/nightingale.wav')
```

We convert to mono by averaging the left and right channels.

```
audio = np.mean(audio, axis=1)
```

Then, we calculate the length of the snippet and plot the audio (Figure 4-2).

```
N = audio.shape[0]
L = N / rate

print(f'Audio length: {L:.2f} seconds')

f, ax = plt.subplots()
ax.plot(np.arange(N) / rate, audio)
ax.set_xlabel('Time [s]')
ax.set_ylabel('Amplitude [unknown]');

Audio length: 7.67 seconds
```

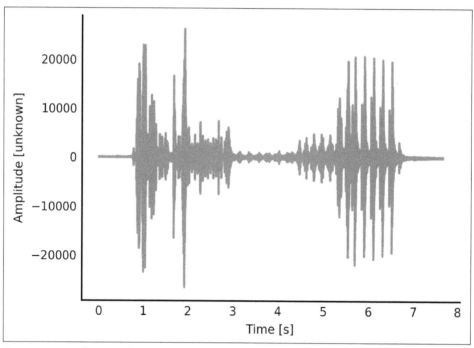

Figure 4-2. Audio waveform plot of nightingale birdsong

Well, that's not very satisfying, is it? If I sent this voltage to a speaker, I might hear a bird chirping, but I can't very well imagine how it would sound in my head. Is there a better way of *seeing* what is going on?

There is, and it is called the discrete Fourier transform, or DFT, where *discrete* refers to the recording consisting of time-spaced sound measurements, in contrast to a continual recording as, for example, on magnetic tape (remember cassettes?). The DFT is often computed using the FFT algorithm, a name informally used to refer to the DFT itself. The DFT tells us which frequencies or "notes" to expect in our signal.

Of course, a bird sings many notes throughout the song, so we'd also like to know *when* each note occurs. The Fourier transform takes a signal in the time domain (i.e., a set of measurements over time) and turns it into a spectrum—a set of frequencies with corresponding (complex[2]) values. The spectrum does not contain any information about time![3]

So, to find both the frequencies and the time at which they were sung, we'll need to be somewhat clever. Our strategy is as follows: take the audio signal, split it into small, overlapping slices, and apply the Fourier transform to each (a technique known as the *short time Fourier transform*).

We'll split the signal into slices of 1,024 samples—that's about 0.02 seconds of audio. The reason we chose 1,024 and not 1,000 we'll explain in a second when we examine performance. The slices will overlap by 100 samples as shown here:

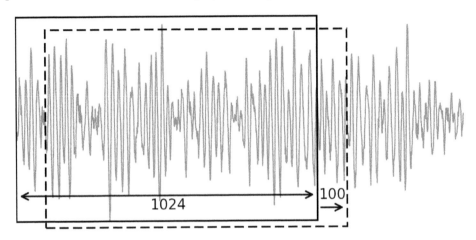

Start by chopping up the signal into slices of 1024 samples, each slice overlapping the previous by 100 samples. The resulting `slices` object contains one slice per row.

```
from skimage import util

M = 1024

slices = util.view_as_windows(audio, window_shape=(M,), step=100)
print(f'Audio shape: {audio.shape}, Sliced audio shape: {slices.shape}')

Audio shape: (338081,), Sliced audio shape: (3371, 1024)
```

---

2 The Fourier transform essentially tells us how to combine a set of sinusoids of varying frequency to form the input signal. The spectrum consists of complex numbers—one for each sinusoid. A complex number encodes two things: a magnitude and an angle. The magnitude is the strength of the sinusoid in the signal, and the angle is how much it is shifted in time. At this point, we only care about the magnitude, which we calculate using np.abs.

3 For more on techniques for calculating both (approximate) frequencies and time of occurrence, read up on wavelet analysis.

Generate a windowing function (see "Windowing" on page 100 for a discussion of the underlying assumptions and interpretations of each) and multiply it with the signal:

```
win = np.hanning(M + 1)[:-1]
slices = slices * win
```

It's more convenient to have one slice per column, so we take the transpose:

```
slices = slices.T
print('Shape of `slices`:', slices.shape)

Shape of `slices`: (1024, 3371)
```

For each slice, calculate the DFT, which returns both positive and negative frequencies (more on that in "Frequencies and Their Ordering" on page 94), so we slice out the positive M2 frequencies for now.

```
spectrum = np.fft.fft(slices, axis=0)[:M // 2 + 1:-1]
spectrum = np.abs(spectrum)
```

(As a quick aside, you'll note that we use `scipy.fftpack.fft` and `np.fft` interchangeably. NumPy provides basic FFT functionality, which SciPy extends further, but both include an `fft` function, based on the Fortran FFTPACK.)

The spectrum can contain both very large and very small values. Taking the log compresses the range significantly.

Here we do a log plot of the ratio of the signal divided by the maximum signal (shown in Figure 4-3). The specific unit used for the ratio is the decibel, $20 log_{10}$ (amplitude ratio).

```
f, ax = plt.subplots(figsize=(4.8, 2.4))

S = np.abs(spectrum)
S = 20 * np.log10(S / np.max(S))

ax.imshow(S, origin='lower', cmap='viridis',
          extent=(0, L, 0, rate / 2 / 1000))
ax.axis('tight')
ax.set_ylabel('Frequency [kHz]')
ax.set_xlabel('Time [s]');
```

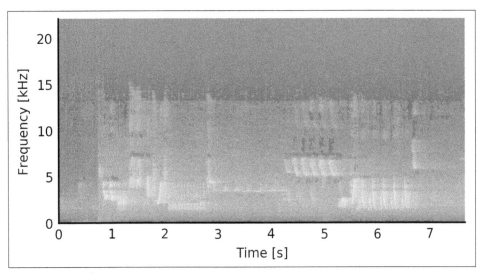

Figure 4-3. Birdsong spectrogram

Much better! We can now see that the frequencies vary over time, and the spectrogram corresponds to the way the audio sounds. See if you can match our earlier description: chee-chee-woorrrr-hee-hee cheet-wheet-hoorrr-chirrr-whi-wheo-wheo-wheo-wheo-wheo-wheo. (I didn't transcribe the 3 to 5 second mark—that's another bird.)

SciPy already includes an implementation of this procedure as scipy.signal.spec trogram (Figure 4-4), which can be invoked as follows:

```
from scipy import signal

freqs, times, Sx = signal.spectrogram(audio, fs=rate, window='hanning',
                                       nperseg=1024, noverlap=M - 100,
                                       detrend=False, scaling='spectrum')

f, ax = plt.subplots(figsize=(4.8, 2.4))
ax.pcolormesh(times, freqs / 1000, 10 * np.log10(Sx), cmap='viridis')
ax.set_ylabel('Frequency [kHz]')
ax.set_xlabel('Time [s]');
```

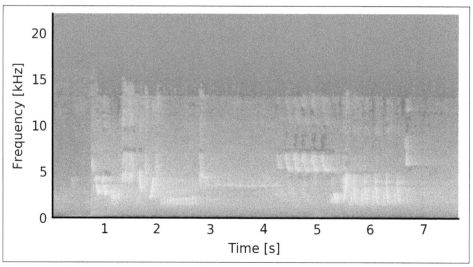

Figure 4-4. SciPy built-in rendition of birdsong spectrogram

The only differences between the manual spectrogram that we created versus the Sci-Py's built-in function are that SciPy returns the spectrum magnitude squared (which turns measured voltage into measured energy), and multiplies it by some normalization factors.[4]

History

Tracing the exact origins of the Fourier transform is tricky. Some related procedures go as far back as Babylonian times, but it was the hot topics of calculating asteroid orbits and solving the heat (flow) equation that led to several breakthroughs in the early 1800s. Whom exactly among Clairaut, Lagrange, Euler, Gauss, and D'Alembert we should thank is not exactly clear, but Gauss was the first to describe the fast Fourier transform (an algorithm for computing the DFT, popularized by Cooley and Tukey in 1965). Joseph Fourier, after whom the transform is named, first claimed that *arbitrary* periodic[5] functions can be expressed as a sum of trigonometric functions.

---

4 SciPy goes to some effort to preserve the energy in the spectrum. Therefore, when taking only half the components (for N even), it multiplies the remaining components, apart from the first and last components, by two (those two components are "shared" by the two halves of the spectrum). It also normalizes the window by dividing it by its sum.

5 The period can, in fact, also be infinite! The general continuous Fourier transform provides for this possibility. DFTs are generally defined over a finite interval, and this is implicitly the period of the time domain function that is transformed. In other words, if you do the inverse DFT, you *always* get a periodic signal out.

Implementation

The DFT functionality in SciPy lives in the `scipy.fftpack` module. Among other things, it provides the following DFT-related functionality:

`fft, fft2, fftn`
> Compute the DFT using the FFT algorithm in 1, 2, or n dimensions.

`ifft, ifft2, ifftn`
> Compute the inverse of the DFT.

`dct, idct, dst, idst`
> Compute the cosine and sine transforms, and their inverses.

`fftshift, ifftshift`
> Shift the zero-frequency component to the center of the spectrum and back, respectively (more about that soon).

`fftfreq`
> Return the DFT sample frequencies.

`rfft`
> Compute the DFT of a real sequence, exploiting the symmetry of the resulting spectrum for increased performance. Also used by `fft` internally when applicable.

This list is complemented by the following functions in NumPy:

`np.hanning, np.hamming, np.bartlett, np.blackman, np.kaiser`
> Tapered windowing functions.

The DFT is also used to perform fast convolutions of large inputs by `scipy.signal.fftconvolve`.

SciPy wraps the Fortran FFTPACK library—it is not the fastest out there, but unlike packages such as FFTW, it has a permissive free software license.

Choosing the Length of the DFT

A naive calculation of the DFT takes $\mathcal{O}(N^2)$ operations.[6] How come? Well, you have N (complex) sinusoids of different frequencies ($2\pi f \times 0$, $2\pi f \times 1$; $2\pi f \times 3$, ..., $2\pi f \times (N - 1)$), and you want to see how strongly your signal corresponds to each. Starting with the first, you take the dot-product with the signal (which, in itself, entails N multiplication operations). Repeating this operation N times, once for each sinusoid, then gives N^2 operations.

Now, contrast that with the FFT, which is $\mathcal{O}(N \log N)$ in the ideal case due to the clever reuse of calculations—a great improvement! However, the classical Cooley-Tukey algorithm implemented in FFTPACK (and used by SciPy) recursively breaks up the transform into smaller (prime-sized) pieces and only shows this improvement for "smooth" input lengths (an input length is considered smooth when its largest prime factor is small, as shown in Figure 4-5). For large prime-sized pieces, the Bluestein or Rader algorithms can be used in conjunction with the Cooley-Tukey algorithm, but this optimization is not implemented in FFTPACK.[7]

Let us illustrate:

```
import time

from scipy import fftpack
from sympy import factorint

K = 1000
lengths = range(250, 260)

# Calculate the smoothness for all input lengths
smoothness = [max(factorint(i).keys()) for i in lengths]


exec_times = []
for i in lengths:
    z = np.random.random(i)
```

---

6 In computer science, the computational cost of an algorithm is often expressed in "Big O" notation. The notation gives us an indication of how an algorithm's execution time scales with an increasing number of elements. If an algorithm is $\mathcal{O}(N)$, its execution time increases linearly with the number of input elements (e.g., searching for a given value in an unsorted list is $\mathcal{O}(N)$. Bubble sort is an example of an $O(N^2)$ algorithm; the exact number of operations performed may, hypothetically, be $N + \frac{1}{2}N^2$, meaning that the computational cost grows quadratically with the number of input elements.

7 While ideally we don't want to reimplement existing algorithms, sometimes it becomes necessary in order to obtain the best execution speeds possible, and tools like Cython (*http://cython.org*), which compiles Python to C, and Numba (*http://numba.pydata.org*), which does just-in-time compilation of Python code, make life a lot easier (and faster!). If you are able to use GPL-licensed software, you may consider using PyFFTW (*https://github.com/hgomersall/pyFFTW*) for faster FFTs.

```
# For each input length i, execute the FFT K times
# and store the execution time

times = []
for k in range(K):
    tic = time.monotonic()
    fftpack.fft(z)
    toc = time.monotonic()
    times.append(toc - tic)

# For each input length, remember the *minimum* execution time
exec_times.append(min(times))


f, (ax0, ax1) = plt.subplots(2, 1, sharex=True)
ax0.stem(lengths, np.array(exec_times) * 10**6)
ax0.set_ylabel('Execution time (μs)')

ax1.stem(lengths, smoothness)
ax1.set_ylabel('Smoothness of input length\n(lower is better)')
ax1.set_xlabel('Length of input');
```

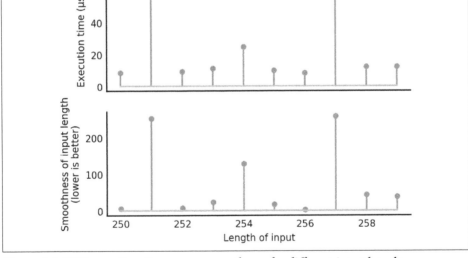

Figure 4-5. FFT execution time versus smoothness for different input lengths

The intuition is that, for smooth numbers, the FFT can be broken up into many small pieces. After performing the FFT on the first piece, we can reuse those results in subsequent computations. This explains why we chose a length of 1,024 for our audio slices earlier—it has a smoothness of only 2, resulting in the optimal "radix-2 Cooley-Tukey" algorithm, which computes the FFT using only $(N/2)\log_2 N = 5,120$ complex multiplications, instead of $N^2 = 1,048,576$. Choosing $N = 2^m$ always ensures a maximally smooth N (and, thus, the fastest FFT).

More DFT Concepts

Next, we present a couple of common concepts worth knowing before operating heavy Fourier transform machinery, whereafter we tackle another real-world problem: analyzing target detection in radar data.

Frequencies and Their Ordering

For historical reasons, most implementations return an array where frequencies vary from low to high to low (see "Discrete Fourier Transforms" on page 95 for further explanation of frequencies). For example, when we do the real Fourier transform of a signal of all ones, an input that has no variation and therefore only has the slowest, constant Fourier component (also known as the "DC," or Direct Current, component —just electronics jargon for "mean of the signal"), appearing as the first entry:

```
from scipy import fftpack
N = 10

fftpack.fft(np.ones(N))   # The first component is np.mean(x) * N
array([ 10.+0.j,    0.+0.j,    0.+0.j,    0.+0.j,    0.+0.j,    0.+0.j,
         0.-0.j,    0.-0.j,    0.-0.j,    0.-0.j])
```

When we try the FFT on a rapidly changing signal, we see a high-frequency component appear:

```
z = np.ones(10)
z[::2] = -1

print(f'Applying FFT to {z}')
fftpack.fft(z)

Applying FFT to [-1.  1. -1.  1. -1.  1. -1.  1. -1.  1.]
array([  0.+0.j,    0.+0.j,    0.+0.j,    0.+0.j,    0.+0.j, -10.+0.j,
         0.-0.j,    0.-0.j,    0.-0.j,    0.-0.j])
```

Note that the FFT returns a complex spectrum that, in the case of real inputs, is conjugate symmetrical (i.e., symmetric in the real part and antisymmetric in the imaginary part):

```
x = np.array([1, 5, 12, 7, 3, 0, 4, 3, 2, 8])
X = fftpack.fft(x)

np.set_printoptions(precision=2)

print("Real part:      ", X.real)
print("Imaginary part:", X.imag)
```

```
np.set_printoptions()
```

```
Real part:      [ 45.      7.09 -12.24  -4.09 -7.76  -1.    -7.76  -4.09 -12.24
                  7.09]
Imaginary part: [  0.    -10.96  -1.62  12.03  6.88   0.    -6.88 -12.03   1.62
                 10.96]
```

(And, again, recall that the first component is np.mean(x) * N.)

The fftfreq function tells us which frequencies we are looking at specifically:

```
fftpack.fftfreq(10)
```

```
array([ 0. ,  0.1,  0.2,  0.3,  0.4, -0.5, -0.4, -0.3, -0.2, -0.1])
```

The result tells us that our maximum component occurred at a frequency of 0.5 cycles per sample. This agrees with the input, where a minus-one-plus-one cycle repeated every second sample.

Sometimes, it is convenient to view the spectrum organized slightly differently, from high-negative to low-to-high-positive (for now, we won't dive too deeply into the concept of negative frequency, other than to say a real-world sine wave is produced by a combination of positive and negative frequencies). We reshuffle the spectrum using the fftshift function.

Discrete Fourier Transforms

The DFT converts a sequence of N equally spaced real or complex samples $x_0, x_1, ... x_{N-1}$ of a function $x(t)$ of time (or another variable, depending on the application) into a sequence of N complex numbers X_k by the following summation:

$$X_k = \sum_{n=0}^{N-1} x_n e^{-j2\pi kn/N}, \; k = 0, 1, ... > N - 1$$

With the numbers X_k known, the inverse DFT *exactly* recovers the sample values x_n through the following summation:

$$x_n = \frac{1}{N} \sum_{k=0}^{N-1} X_k e^{j2\pi kn/N}$$

Keeping in mind that $e^{j\theta} = cos\theta + j \, sin \, \theta$, the last equation shows that the DFT has decomposed the sequence x_n into a complex discrete Fourier series with coefficients X_k. Comparing the DFT with a continuous complex Fourier series:

$$x(t) = \sum_{n=-\infty}^{\infty} c_n e^{jn\omega_0 t}$$

The DFT is a *finite* series with N terms defined at the equally spaced discrete instances of the *angle* $(\omega_0 t_n) = 2\pi\frac{k}{N}$ in the interval $[0, 2\pi)$—that is, *including* 0 and *excluding* 2π. This automatically normalizes the DFT so that time does not appear explicitly in the forward or inverse transform.

If the original function $x(t)$ is limited in frequency to less than half of the sampling frequency (the so-called *Nyquist frequency*), interpolation between sample values produced by the inverse DFT will usually give a faithful reconstruction of $x(t)$. If $x(t)$ is *not* limited as such, the inverse DFT can, in general, not be used to reconstruct $x(t)$ by interpolation. Note that this limit does not imply that there are *no* methods that can do such a reconstruction—see, for example, compressed sensing, or finite rate of innovation sampling.

The function $e^{j2\pi k/N} = (e^{j2\pi/N})^k = w^k$ takes on discrete values between 0 and $2\pi\frac{N-1}{N}$ on the unit circle in the complex plane. The function $e^{j2\pi kn/N} = w^{kn}$ encircles the origin $n\frac{N-1}{N}$ times, thus generating harmonics of the fundamental sinusoid for which $n = 1$.

The way in which we defined the DFT leads to a few subtleties when $n > \frac{N}{2}$, for even N.[8] The function $e^{j2\pi kn/N}$ is plotted for increasing values of k in Figure 4-6, for the cases $n = 1$ to $n = N - 1$ for $N = 16$. When k increases from k to $k + 1$, the angle increases by $\frac{2\pi n}{N}$. When $n = 1$, the step is $\frac{2\pi}{N}$. When $n = N - 1$, the angle increases by $2\pi\frac{N-1}{N} = 2\pi - \frac{2\pi}{N}$. Since 2π is precisely once around the circle, the step equates to $-\frac{2\pi}{N}$ —that is, in the direction of a negative frequency. The components up to $N/2$ represent *positive* frequency components, those above $N/2$ up to $N - 1$ represent *negative* frequencies. The angle increment for the component $N/2$ for N even advances precisely halfway around the circle for each increment in k and can therefore be interpreted as either a positive or a negative frequency. This component of the DFT represents the Nyquist frequency (i.e., half of the sampling frequency), and is useful to orientate oneself when looking at DFT graphics.

The FFT in turn is simply a special and highly efficient algorithm for calculating the DFT. Whereas a straightforward calculation of the DFT takes of the order of N^2 calculations to compute, the FFT algorithm requires of the order $N \log N$ calculations. The FFT was key to the widespread use of DFT in real-time applications and was included in a list of the top 10 algorithms of the twentieth century by the *IEEE journal Computing in Science & Engineering* in 2000.

---

8 We leave it as an exercise for the reader to picture the situation for N odd. In this chapter, all examples use even-order DFTs.

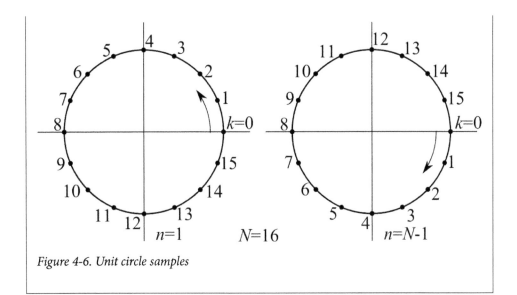

Figure 4-6. Unit circle samples

Let's examine the frequency components in a noisy image (Figure 4-7). Note that, while a static image has no time-varying component, its values do vary across *space*. The DFT applies equally to either case.

First, load and display the image:

```
from skimage import io
image = io.imread('images/moonlanding.png')
M, N = image.shape

f, ax = plt.subplots(figsize=(4.8, 4.8))
ax.imshow(image)

print((M, N), image.dtype)
```

```
(474, 630) uint8
```

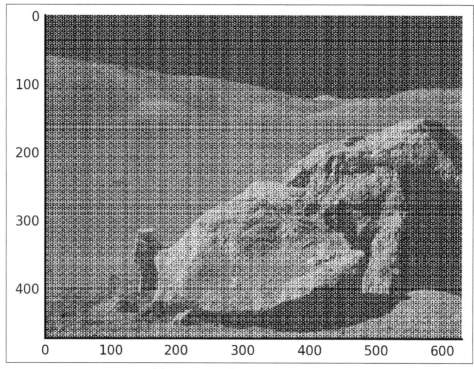

Figure 4-7. A noisy image of the moon landing

Do not adjust your monitor! The image you are seeing is real, although clearly distorted by either the measurement or transmission equipment.

To examine the spectrum of the image, we use `fftn` (instead of `fft`) to compute the DFT, since it has more than one dimension. The 2D FFT is equivalent to taking the 1D FFT across rows and then across columns, or vice versa.

```
F = fftpack.fftn(image)

F_magnitude = np.abs(F)
F_magnitude = fftpack.fftshift(F_magnitude)
```

Again, we take the log of the spectrum to compress the range of values, before displaying:

```
f, ax = plt.subplots(figsize=(4.8, 4.8))

ax.imshow(np.log(1 + F_magnitude), cmap='viridis',
          extent=(-N // 2, N // 2, -M // 2, M // 2))
ax.set_title('Spectrum magnitude');
```

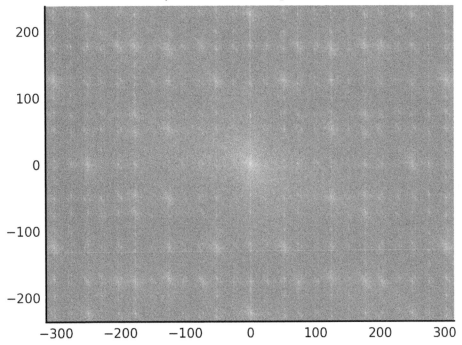

Spectrum magnitude

Note the high values around the origin (middle) of the spectrum—these coefficients describe the low frequencies or smooth parts of the image, a vague canvas of the photo. Higher-frequency components, spread throughout the spectrum, fill in the edges and detail. Peaks around higher frequencies correspond to the periodic noise.

From the photo, we can see that the noise (measurement artifacts) is highly periodic, so we hope to remove it by zeroing out the corresponding parts of the spectrum (Figure 4-8).

The image with those peaks suppressed indeed looks quite different!

```
# Set block around center of spectrum to zero
K = 40
F_magnitude[M // 2 - K: M // 2 + K, N // 2 - K: N // 2 + K] = 0

# Find all peaks higher than the 98th percentile
peaks = F_magnitude < np.percentile(F_magnitude, 98)

# Shift the peaks back to align with the original spectrum
peaks = fftpack.ifftshift(peaks)

# Make a copy of the original (complex) spectrum
F_dim = F.copy()
```

```
# Set those peak coefficients to zero
F_dim = F_dim * peaks.astype(int)

# Do the inverse Fourier transform to get back to an image.
# Since we started with a real image, we only look at the real part of
# the output.
image_filtered = np.real(fftpack.ifft2(F_dim))

f, (ax0, ax1) = plt.subplots(2, 1, figsize=(4.8, 7))
ax0.imshow(np.log10(1 + np.abs(F_dim)), cmap='viridis')
ax0.set_title('Spectrum after suppression')

ax1.imshow(image_filtered)
ax1.set_title('Reconstructed image');
```

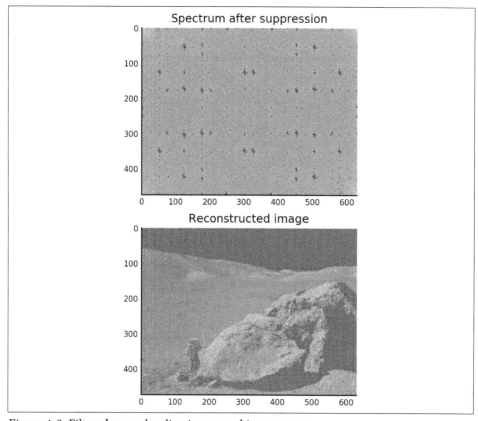

Figure 4-8. Filtered moon landing image and its spectrum

Windowing

If we examine the Fourier transform of a rectangular pulse, we see significant side lobes in the spectrum:

```
x = np.zeros(500)
x[100:150] = 1

X = fftpack.fft(x)

f, (ax0, ax1) = plt.subplots(2, 1, sharex=True)

ax0.plot(x)
ax0.set_ylim(-0.1, 1.1)

ax1.plot(fftpack.fftshift(np.abs(X)))
ax1.set_ylim(-5, 55);
```

In theory, you would need a combination of infinitely many sinusoids (frequencies) to represent any abrupt transition; the coefficients would typically have the same side lobe structure as seen here for the pulse.

Importantly, the DFT assumes that the input signal is periodic. If the signal is not, the assumption is simply that, right at the end of the signal, it jumps back to its beginning value. Consider the function, $x(t)$, shown here:

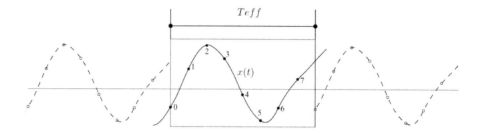

We measure the signal for only a short time, labeled T_{eff}. The Fourier transform assumes that $x(8) = x(0)$, and that the signal is continued as the dashed, rather than the solid, line. This introduces a big jump at the edge, with the expected oscillation in the spectrum:

```python
t = np.linspace(0, 1, 500)
x = np.sin(49 * np.pi * t)

X = fftpack.fft(x)

f, (ax0, ax1) = plt.subplots(2, 1)

ax0.plot(x)
ax0.set_ylim(-1.1, 1.1)

ax1.plot(fftpack.fftfreq(len(t)), np.abs(X))
ax1.set_ylim(0, 190);
```

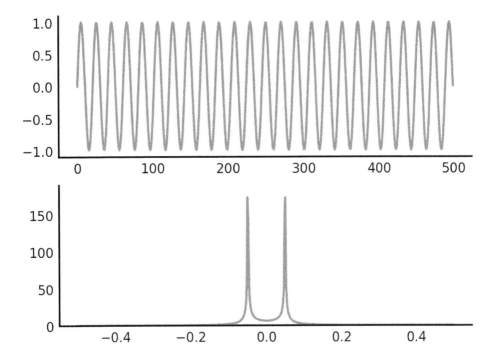

Instead of the expected two lines, the peaks are spread out in the spectrum.

We can counter this effect by a process called *windowing*. The original function is multiplied with a window function such as the Kaiser window $K(N, \beta)$. Here we visualize it for β ranging from 0 to 100:

```
f, ax = plt.subplots()

N = 10
beta_max = 5
colormap = plt.cm.plasma

norm = plt.Normalize(vmin=0, vmax=beta_max)

lines = [
    ax.plot(np.kaiser(100, beta), color=colormap(norm(beta)))
    for beta in np.linspace(0, beta_max, N)
    ]

sm = plt.cm.ScalarMappable(cmap=colormap, norm=norm)

sm._A = []

plt.colorbar(sm).set_label(r'Kaiser $\beta$');
```

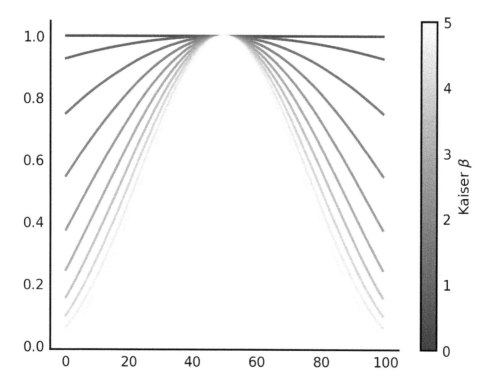

By changing the parameter β, we can change the shape of the window from rectangular ($\beta = 0$, no windowing) to a window that produces signals that smoothly increase from zero and decrease to zero at the endpoints of the sampled interval, producing very low side lobes (β typically between 5 and 10).[9]

Applying the Kaiser window here, we see that the side lobes have been drastically reduced, at the cost of a slight widening in the main lobe.

The effect of windowing our previous example is noticeable:

```
win = np.kaiser(len(t), 5)
X_win = fftpack.fft(x * win)

plt.plot(fftpack.fftfreq(len(t)), np.abs(X_win))
plt.ylim(0, 190);
```

---

9 The classical windowing functions include Hann, Hamming, and Blackman. They differ in their side lobe levels and in the broadening of the main lobe (in the Fourier domain). A modern and flexible window function that is close to optimal for most applications is the Kaiser window—a good approximation to the optimal prolate spheroid window, which concentrates the most energy into the main lobe. We can tune the Kaiser window to suit the particular application, as illustrated in the main text, by adjusting the parameter β.

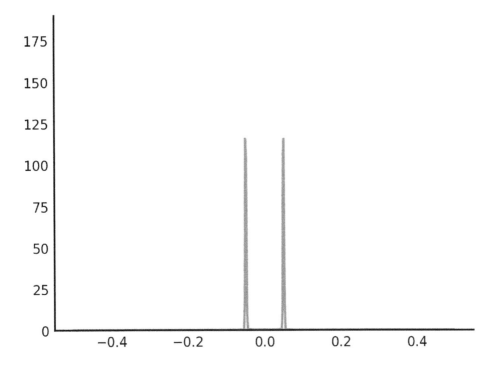

Real-World Application: Analyzing Radar Data

Linearly modulated FMCW (Frequency-Modulated Continuous-Wave) radars make extensive use of the FFT algorithm for signal processing and provide examples of various applications of the FFT. We will use actual data from an FMCW radar to demonstrate one such application: target detection.

Roughly, an FMCW radar works like this (see "A Simple FMCW Radar System" on page 106 and Figure 4-9 for more detail):

1. A signal with changing frequency is generated. This signal is transmitted by an antenna, after which it travels outward, away from the radar. When it hits an object, part of the signal is reflected back to the radar, where it is received, multiplied by a copy of the transmitted signal, and sampled, turning it into numbers that are packed into an array. Our challenge is to interpret those numbers to form meaningful results.

2. The preceding multiplication step is important. From school, recall the trigonometric identity:

$$\sin(xt)\sin(yt) = \tfrac{1}{2}\left[\sin\left((x-y)t + \tfrac{\pi}{2}\right) - \sin\left((x+y)t + \tfrac{\pi}{2}\right)\right]$$

3. Thus, if we multiply the received signal by the transmitted signal, we expect two frequency components to appear in the spectrum: one that is the difference in frequencies between the received and transmitted signal, and one that is the sum of their frequencies.

4. We are particularly interested in the first, since that gives us some indication of how long it took the signal to reflect back to the radar (in other words, how far away the object is from us!). We discard the other by applying a low-pass filter to the signal (i.e., a filter that discards any high frequencies).

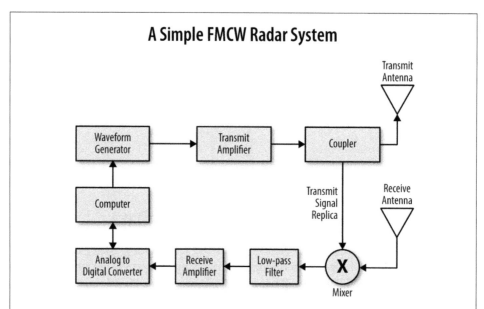

Figure 4-9. The block diagram of a simple FMCW radar system

A block diagram of a simple FMCW radar that uses separate transmit and receive antennas is shown above. The radar consists of a waveform generator that generates a sinusoidal signal of which the frequency varies linearly around the required transmit frequency. The generated signal is amplified to the required power level by the transmit amplifier and routed to the transmit antenna via a coupler circuit where a copy of the transmit signal is tapped off. The transmit antenna radiates the transmit signal as an electromagnetic wave in a narrow beam toward the target to be detected. When the wave encounters an object that reflects electromagnetic waves, a fraction of the energy irradiating the target is reflected back to the receiver as a second electromagnetic wave that propagates in the direction of the radar system. When this wave encounters the receive antenna, the antenna collects the energy in the wave energy impinging on it and converts it to a fluctuating voltage that is fed to the mixer. The mixer multiplies the received signal with a replica of the transmit signal and produces a sinusoidal signal with a frequency equal to the difference in frequency between the transmitted and received signals. The low-pass filter ensures that the received signal is band limited (i.e., does not contain frequencies that we don't care about) and the receive amplifier strengthens the signal to a suitable amplitude for the analog-to-digital converter (ADC) that feeds data to the computer.

To summarize, we should note that:

- The data that reaches the computer consists of N samples sampled (from the multiplied, filtered signal) at a sample frequency of f_s.
- The *amplitude* of the returned signal varies depending on the *strength of the reflection* (i.e., is a property of the target object and the distance between the target and the radar).
- The *frequency measured* is an indication of the *distance* of the target object from the radar.

To start our analysis of radar data, we'll generate some synthetic signals, after which we'll turn our focus to the output of an actual radar.

Recall that the radar is increasing its frequency as it transmits at a rate of S Hz/s. After a certain amount of time, t, has passed, the frequency will now be tS higher (Figure 4-10). In that same time span, the radar signal has traveled $d = t/v$ meters, where v is the speed of the transmitted wave through air (roughly the same as the speed of light, 3×10^8 m/s).

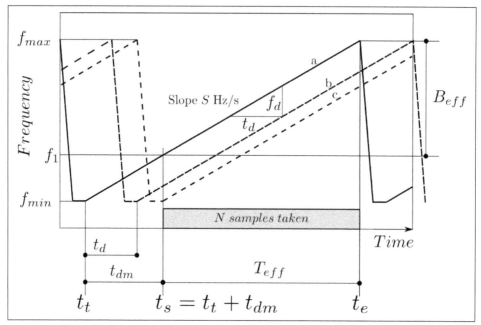

Figure 4-10. The frequency relationships in an FMCW radar with linear frequency modulation

Combining the above observations, we can calculate the amount of time it would take the signal to travel to, bounce off, and return from a target that is distance R away:

$t_R = 2R/v$

```
pi = np.pi

# Radar parameters
fs = 78125          # Sampling frequency in Hz, i.e., we sample 78125
                    # times per second

ts = 1 / fs         # Sampling time, i.e., one sample is taken each
                    # ts seconds

Teff = 2048.0 * ts  # Total sampling time for 2048 samples
                    # (AKA effective sweep duration) in seconds.

Beff = 100e6        # Range of transmit signal frequency during the time the
                    # radar samples, known as the "effective bandwidth"
                    # (given in Hz)

S = Beff / Teff     # Frequency sweep rate in Hz/s

# Specification of targets.  We made these targets up, imagining they
```

```
# are objects seen by the radar with the specified range and size.

R = np.array([100, 137, 154, 159,  180])  # Ranges (in meter)
M = np.array([0.33, 0.2, 0.9, 0.02, 0.1])  # Target size
P = np.array([0, pi / 2, pi / 3, pi / 5, pi / 6])  # Randomly chosen phase offsets

t = np.arange(2048) * ts  # Sample times

fd = 2 * S * R / 3E8      # Frequency differences for these targets

# Generate five targets
signals = np.cos(2 * pi * fd * t[:, np.newaxis] + P)

# Save the signal associated with the first target as an example for
# later inspection
v_single = signals[:, 0]

# Weigh the signals, according to target size and sum, to generate
# the combined signal seen by the radar.
v_sim = np.sum(M * signals, axis=1)

## The above code is equivalent to:
#
# v0 = np.cos(2 * pi * fd[0] * t)
# v1 = np.cos(2 * pi * fd[1] * t + pi / 2)
# v2 = np.cos(2 * pi * fd[2] * t + pi / 3)
# v3 = np.cos(2 * pi * fd[3] * t + pi / 5)
# v4 = np.cos(2 * pi * fd[4] * t + pi / 6)
#
## Blend them together
# v_single = v0
# v_sim = (0.33 * v0) + (0.2 * v1) + (0.9 * v2) + (0.02 * v3) + (0.1 * v4)
```

Here, we generated a synthetic signal, v_{single}, received when looking at a single target (see Figure 4-11). By counting the number of cycles seen in a given time period, we can compute the frequency of the signal and thus the distance to the target.

A real radar will rarely receive only a single echo, though. The simulated signal v_{sim} shows what a radar signal will look like with five targets at different ranges (including two close to one another at 154 and 159 meters), and $v_{actual}(t)$ shows the output signal obtained with an actual radar. When we add multiple echoes together, the result makes little intuitive sense (Figure 4-11); until, that is, we look at it more carefully through the lens of the DFT.

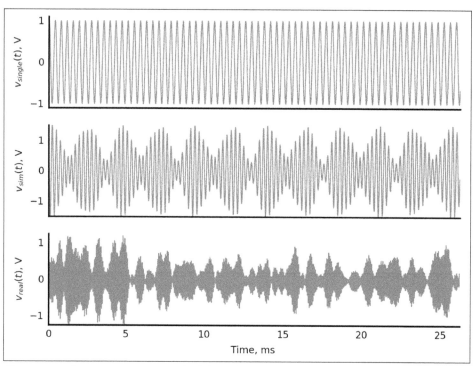

Figure 4-11. Receiver output signals: (a) single simulated target, (b) five simulated targets, and (c) actual radar data

The real-world radar data is read from a NumPy-format *.npz* file (a lightweight, cross-platform, and cross-version compatible storage format). These files can be saved with the np.savez or np.savez_compressed functions. Note that SciPy's io submodule can also easily read other formats, such as MATLAB and NetCDF files.

```
data = np.load('data/radar_scan_0.npz')

# Load variable 'scan' from 'radar_scan_0.npz'
scan = data['scan']

# The dataset contains multiple measurements, each taken with the
# radar pointing in a different direction.  Here we take one such as
# measurement, at a specified azimuth (left-right position) and elevation
# (up-down position).  The measurement has shape (2048,).

v_actual = scan['samples'][5, 14, :]

# The signal amplitude ranges from -2.5V to +2.5V.  The 14-bit
# analogue-to-digital converter in the radar gives out integers
# between -8192 to 8192.  We convert back to voltage by multiplying by
# $(2.5 / 8192)$.
```

```
v_actual = v_actual * (2.5 / 8192)
```

Since *.npz* files can store multiple variables, we have to select the one we want: data['scan']. That returns a *structured NumPy array* with the following fields:

time
> Unsigned 64-bit (8 byte) integer (np.uint64)

size
> Unsigned 32-bit (4 byte) integer (np.uint32)

position

> az
> > 32-bit float (np.float32)

> el
> > 32-bit float (np.float32)

> region_type
> > Unsigned 8-bit (1 byte) integer (np.uint8)

> region_ID
> > Unsigned 16-bit (2 byte) integer (np.uint16)

> gain
> > Unsigned 8-bit (1 byte) integer (np.uin8)

> samples
> > 2,048 unsigned 16-bit (2 byte) integers (np.uint16)

While it is true that NumPy arrays are *homogeneous* (i.e., all the elements inside are the same), it does not mean that those elements cannot be compound elements, as is the case here.

An individual field is accessed using dictionary syntax:

```
azimuths = scan['position']['az']  # Get all azimuth measurements
```

To summarize what we've seen so far: the shown measurements (v_{sim} and v_{actual}) are the sum of sinusoidal signals reflected by each of several objects. We need to determine each of the constituent components of these composite radar signals. The FFT is the tool that will do this for us.

Signal Properties in the Frequency Domain

First, we take the FFTs of our three signals (synthetic single target, synthetic multi-target, and real) and then display the positive frequency components (i.e., components 0 to *N/2*; see Figure 4-12). These are called the *range traces* in radar terminology.

```
fig, axes = plt.subplots(3, 1, sharex=True, figsize=(4.8, 2.4))

# Take FFTs of our signals.  Note the convention to name FFTs with a
# capital letter.

V_single = np.fft.fft(v_single)
V_sim = np.fft.fft(v_sim)
V_actual = np.fft.fft(v_actual)

N = len(V_single)

with plt.style.context('style/thinner.mplstyle'):
    axes[0].plot(np.abs(V_single[:N // 2]))
    axes[0].set_ylabel("$|V_\mathrm{single}|$")
    axes[0].set_xlim(0, N // 2)
    axes[0].set_ylim(0, 1100)

    axes[1].plot(np.abs(V_sim[:N // 2]))
    axes[1].set_ylabel("$|V_\mathrm{sim} |$")
    axes[1].set_ylim(0, 1000)

    axes[2].plot(np.abs(V_actual[:N // 2]))
    axes[2].set_ylim(0, 750)
    axes[2].set_ylabel("$|V_\mathrm{actual}|$")

    axes[2].set_xlabel("FFT component $n$")

    for ax in axes:
        ax.grid()
```

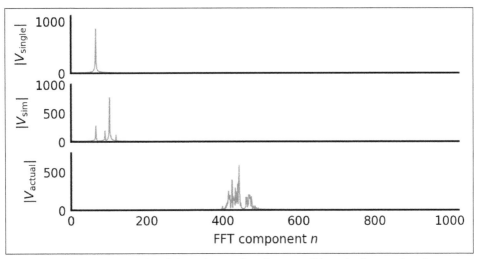

Figure 4-12. Range traces for: (a) single simulated target, (b) multiple simulated targets, and (c) real-world targets

Suddenly, the information makes sense!

The plot for $|V_0|$ clearly shows a target at component 67, and for $|V_{sim}|$ shows the targets that produced the signal that was uninterpretable in the time domain. The real radar signal, $|V_{actual}|$, shows a large number of targets between component 400 and 500 with a large peak in component 443. This happens to be an echo return from a radar illuminating the high wall of an open-cast mine.

To get useful information from the plot, we must determine the range! Again, we use the formula:

$$R_n = \frac{nv}{2B_{eff}}$$

In radar terminology, each DFT component is known as a *range bin*.

This equation also defines the range resolution of the radar: targets will only be distinguishable if they are separated by more than two range bins, for example:

$$\Delta R > \frac{1}{B_{eff}}$$

This is a fundamental property of all types of radar.

This result is quite satisfying—but the dynamic range is so large that we could very easily miss some peaks. Let's take the log as before with the spectrogram:

```
c = 3e8  # Approximately the speed of light and of
         # electromagnetic waves in air

fig, (ax0, ax1, ax2) = plt.subplots(3, 1)


def dB(y):
    "Calculate the log ratio of y / max(y) in decibel."

    y = np.abs(y)
    y /= y.max()

    return 20 * np.log10(y)


def log_plot_normalized(x, y, ylabel, ax):
    ax.plot(x, dB(y))
    ax.set_ylabel(ylabel)
    ax.grid()


rng = np.arange(N // 2) * c / 2 / Beff

with plt.style.context('style/thinner.mplstyle'):
    log_plot_normalized(rng, V_single[:N // 2], "$|V_0|$ [dB]", ax0)
    log_plot_normalized(rng, V_sim[:N // 2], "$|V_5|$ [dB]", ax1)
    log_plot_normalized(rng, V_actual[:N // 2], "$|V_{\mathrm{actual}}|$ [dB]"
        , ax2)

ax0.set_xlim(0, 300)  # Change x limits for these plots so that
ax1.set_xlim(0, 300)  # we are better able to see the shape of the peaks.
ax2.set_xlim(0, len(V_actual) // 2)
ax2.set_xlabel('range')
```

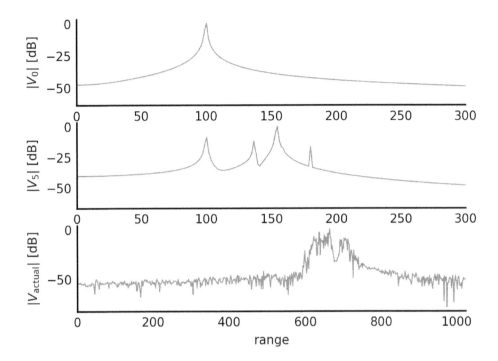

The observable dynamic range is much improved in these plots. For instance, in the real radar signal the *noise floor* of the radar has become visible (i.e., the level where electronic noise in the system starts to limit the radar's ability to detect a target).

Windowing, Applied

We're getting there, but in the spectrum of the simulated signal, we still cannot distinguish the peaks at 154 and 159 meters. Who knows what we're missing in the real-world signal! To sharpen the peaks, we'll return to our toolbox and make use of *windowing*.

Here are the signals used thus far in this example, windowed with a Kaiser window with $\beta = 6.1$:

```
f, axes = plt.subplots(3, 1, sharex=True, figsize=(4.8, 2.8))

t_ms = t * 1000  # Sample times in milli-second

w = np.kaiser(N, 6.1)  # Kaiser window with beta = 6.1

for n, (signal, label) in enumerate([(v_single, r'$v_0 [V]$'),
                                     (v_sim, r'$v_5 [V]$'),
                                     (v_actual, r'$v_{\mathrm{actual}} [V]$')]):
    with plt.style.context('style/thinner.mplstyle'):
```

```
    axes[n].plot(t_ms, w * signal)
    axes[n].set_ylabel(label)
    axes[n].grid()

axes[2].set_xlim(0, t_ms[-1])
axes[2].set_xlabel('Time [ms]');
```

And the corresponding FFTs, or "range traces," in radar terms:

```
V_single_win = np.fft.fft(w * v_single)
V_sim_win = np.fft.fft(w * v_sim)
V_actual_win = np.fft.fft(w * v_actual)

fig, (ax0, ax1,ax2) = plt.subplots(3, 1)

with plt.style.context('style/thinner.mplstyle'):
    log_plot_normalized(rng, V_single_win[:N // 2],
                        r"$|V_{0,\mathrm{win}}|$ [dB]", ax0)
    log_plot_normalized(rng, V_sim_win[:N // 2],
                        r"$|V_{5,\mathrm{win}}|$ [dB]", ax1)
    log_plot_normalized(rng, V_actual_win[:N // 2],
                        r"$|V_\mathrm{actual,win}|$ [dB]", ax2)

ax0.set_xlim(0, 300)  # Change x limits for these plots so that
ax1.set_xlim(0, 300)  # we are better able to see the shape of the peaks.

ax1.annotate("New, previously unseen!", (160, -35), xytext=(10, 15),
             textcoords="offset points", color='red', size='x-small',
             arrowprops=dict(width=0.5, headwidth=3, headlength=4,
                             fc='k', shrink=0.1));
```

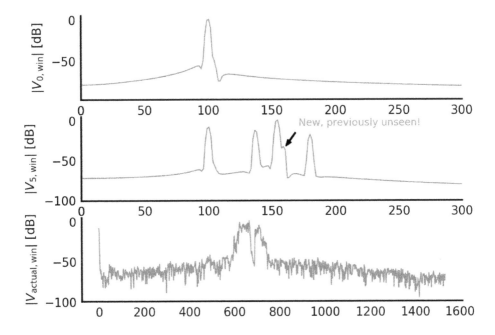

New, previously unseen!

Compare these with the earlier range traces. There is a dramatic lowering in side-lobe level, but at a price: the peaks have changed in shape, widening and becoming less peaky, thus lowering the radar resolution—that is, the ability of the radar to distinguish between two closely spaced targets. The choice of window is a compromise between side-lobe level and resolution. Even so, referring to the trace for V_{sim}, windowing has dramatically increased our ability to distinguish the small target from its large neighbor.

In the real radar data range trace, windowing has also reduced the side lobes. This is most visible in the depth of the notch between the two groups of targets.

Radar Images

Knowing how to analyze a single trace, we can expand to looking at radar images.

The data is produced by a radar with a parabolic reflector antenna. It produces a highly directive round pencil beam with a two-degree spreading angle between half-power points. When directed with normal incidence at a plane, the radar will illuminate a spot of about 2m in diameter at a distance of 60m. Outside this spot, the power drops off quite rapidly, but strong echoes from outside the spot will nevertheless still be visible.

By varying the pencil beam's azimuth (left-right position) and elevation (up-down position), we can sweep it across the target area of interest. When reflections are picked up, we can calculate the distance to the reflector (the object hit by the radar signal). Together with the current pencil beam azimuth and elevation, this defines the reflector's position in 3D.

A rock slope consists of thousands of reflectors. A range bin can be thought of as a large sphere with the radar at its center that intersects the slope along a ragged line. The scatterers on this line will produce reflections in this range bin. The wavelength of the radar (distance the transmitted wave travels in one oscillation second) is about 30mm. The reflections from scatterers separated by odd multiples of a quarter wavelength in range, about 7.5mm, will tend to interfere destructively, while those from scatterers separated by multiples of a half wavelength will tend to interfere constructively at the radar. The reflections combine to produce apparent spots of strong reflections. This specific radar moves its antenna in order to scan small regions consisting of 20 degrees azimuth and 30 degrees elevation bins scanned in steps of 0.5 degrees.

We will now draw some contour plots of the resulting radar data. Refer to Figure 4-13 to see how the different slices are taken. A first slice at fixed range shows the strength of echoes against elevation and azimuth. Another two slices at fixed elevation and azimuth, respectively, show the slope (see Figures 4-13 and 4-14). The stepped construction of the high wall in an opencast mine is visible in the azimuth plane.

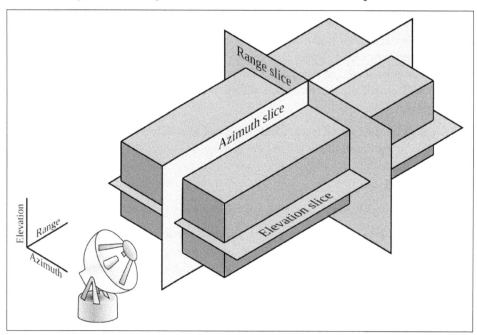

Figure 4-13. Diagram showing azimuth, elevation, and range slices through data volume

```
data = np.load('data/radar_scan_1.npz')
scan = data['scan']

# The signal amplitude ranges from -2.5V to +2.5V.  The 14-bit
# analogue-to-digital converter in the radar gives out integers
# between -8192 to 8192.  We convert back to voltage by multiplying by
# $(2.5 / 8192)$.

v = scan['samples'] * 2.5 / 8192
win = np.hanning(N + 1)[:-1]

# Take FFT for each measurement
V = np.fft.fft(v * win, axis=2)[::-1, :, :N // 2]

contours = np.arange(-40, 1, 2)

# ignore MPL layout warnings
import warnings
warnings.filterwarnings('ignore', '.*Axes.*compatible.*tight_layout.*')

f, axes = plt.subplots(2, 2, figsize=(4.8, 4.8), tight_layout=True)

labels = ('Range', 'Azimuth', 'Elevation')

def plot_slice(ax, radar_slice, title, xlabel, ylabel):
    ax.contourf(dB(radar_slice), contours, cmap='magma_r')
    ax.set_title(title)
    ax.set_xlabel(xlabel)
    ax.set_ylabel(ylabel)
    ax.set_facecolor(plt.cm.magma_r(-40))

with plt.style.context('style/thinner.mplstyle'):
    plot_slice(axes[0, 0], V[:, :, 250], 'Range=250', 'Azimuth', 'Elevation')
    plot_slice(axes[0, 1], V[:, 3, :], 'Azimuth=3', 'Range', 'Elevation')
    plot_slice(axes[1, 0], V[6, :, :].T, 'Elevation=6', 'Azimuth', 'Range')
    axes[1, 1].axis('off')
```

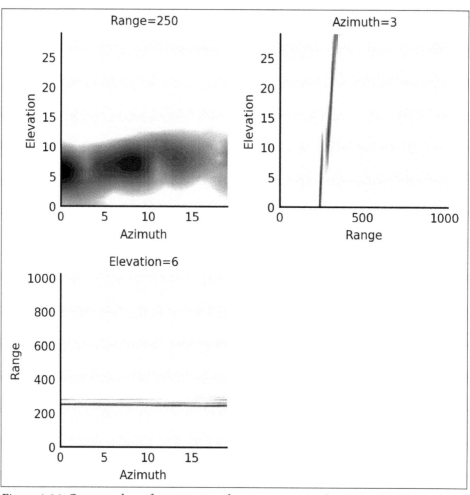

Figure 4-14. Contour plots of range traces along various axes (see Figure 4-13)

3D visualization

We can also visualize the volume in three dimensions (Figure 4-15).

We first compute the argmax (the index of the maximum value) in the range direction. This should give an indication of the range at which the radar beam hit the rock slope. Each argmax index is converted to a 3D (elevation-azimuth-range) coordinate:

```
r = np.argmax(V, axis=2)

el, az = np.meshgrid(*[np.arange(s) for s in r.shape], indexing='ij')

axis_labels = ['Elevation', 'Azimuth', 'Range']
coords = np.column_stack((el.flat, az.flat, r.flat))
```

Taking these coordinates, we project them onto the azimuth-elevation plane (by dropping the range coordinate) and perform a Delaunay tesselation. The tesselation returns a set of indices into our coordinates that define triangles (or simplices). While the triangles are strictly speaking defined on the projected coordinates, we use our original coordinates for the reconstruction, thereby adding back the range component:

```
from scipy import spatial

d = spatial.Delaunay(coords[:, :2])
simplexes = coords[d.vertices]
```

For display purposes, we swap the range axis to be the first:

```
coords = np.roll(coords, shift=-1, axis=1)
axis_labels = np.roll(axis_labels, shift=-1)
```

Now, Matplotlib's `trisurf` can be used to visualize the result:

```
# This import initializes Matplotlib's 3D machinery
from mpl_toolkits.mplot3d import Axes3D

# Set up the 3D axis
f, ax = plt.subplots(1, 1, figsize=(4.8, 4.8),
                     subplot_kw=dict(projection='3d'))

with plt.style.context('style/thinner.mplstyle'):
    ax.plot_trisurf(*coords.T, triangles=d.vertices, cmap='magma_r')

    ax.set_xlabel(axis_labels[0])
    ax.set_ylabel(axis_labels[1])
    ax.set_zlabel(axis_labels[2], labelpad=-3)
    ax.set_xticks([0, 5, 10, 15])

# Adjust the camera position to match our diagram above
ax.view_init(azim=-50);
```

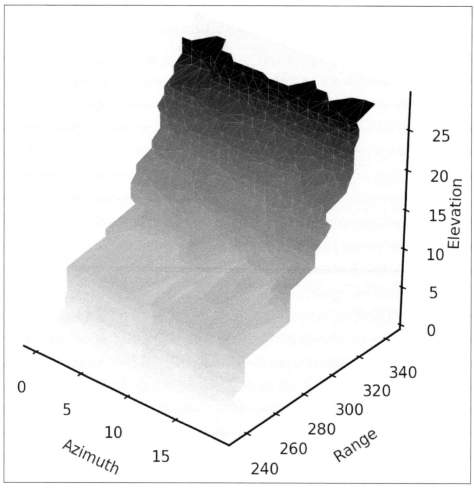

Figure 4-15. 3D visualization of estimated rock slope position

Further Applications of the FFT

The preceding examples show just one of the uses of the FFT in radar. There are many others, such as movement (Doppler) measurement and target recognition. The FFT is pervasive, and is seen everywhere from MRI to statistics. With the basic techniques that this chapter outlines in hand, you should be well equipped to use it!

Further Reading

On the Fourier transform:

- Athanasios Papoulis, *The Fourier Integral and Its Applications* (New York: McGraw-Hill, 1960).

- Ronald A. Bracewell, *The Fourier Transform and Its Applications* (New York: McGraw-Hill, 1986).

On radar signal processing:

- Mark A. Richards, James A. Scheer, and William A. Holm, eds., *Principles of Modern Radar: Basic Principles* (Raleigh, NC: SciTech, 2010).
- Mark A. Richards, *Fundamentals of Radar Signal Processing* (New York: McGraw-Hill, 2014).

Exercise: Image Convolution

The FFT is often used to speed up image convolution (convolution is the application of a sliding filter). Convolve an image with `np.ones((5, 5))`, using a) NumPy's `np.convolve` and b) `np.fft.fft2`. Confirm that the results are identical.

Hints:

- The convolution of x and y is equivalent to `ifft2(X * Y)`, where X and Y are the FFTs of x and y, respectively.
- In order to multiply X and Y, they have to be the same size. Use `np.pad` to extend x and y with zeros (toward the right and bottom) *before* taking their FFT.
- You may see some edge effects. You can remove these by increasing the padding size, so that both x and y have dimensions `shape(x)` + `shape(y)` - `1`.

Check out "Solution: Image Convolution" on page 230.

## CHAPTER 5

Contingency Tables Using Sparse Coordinate Matrices

I like sparseness. There's something about that minimalist feel that can make something have an immediate impact and make it unique. I'll probably always work with that formula. I just don't know how.

    —Britt Daniel, lead singer of *Spoon*

Many real-world matrices are *sparse*, which means that most of their values are zero.

Using NumPy arrays to manipulate sparse matrices wastes a lot of time and energy multiplying many, many values by 0. Instead, we can use SciPy's `sparse` module to solve these efficiently, examining only nonzero values. In addition to helping solve these "canonical" sparse matrix problems, `sparse` can be used for problems that are not obviously related to sparse matrices.

One such problem is the comparison of image segmentations. (Review Chapter 3 for a definition of segmentation.)

The code sample motivating this chapter uses sparse matrices twice. First, we use code nominated by Andreas Mueller to compute a *contingency matrix* that counts the correspondence of labels between two segmentations. Then, with suggestions from Jaime Fernández del Río and Warren Weckesser, we use that contingency matrix to compute the *variation of information*, which measures the differences between segmentations.

```
def variation_of_information(x, y):
    # compute contingency matrix, aka joint probability matrix
    n = x.size
    Pxy = sparse.coo_matrix((np.full(n, 1/n), (x.ravel(), y.ravel())),
                            dtype=float).tocsr()
```

```
# compute marginal probabilities, converting to 1D array
px = np.ravel(Pxy.sum(axis=1))
py = np.ravel(Pxy.sum(axis=0))

# use sparse matrix linear algebra to compute VI
# first, compute the inverse diagonal matrices
Px_inv = sparse.diags(invert_nonzero(px))
Py_inv = sparse.diags(invert_nonzero(py))

# then, compute the entropies
hygx = px @ xlog1x(Px_inv @ Pxy).sum(axis=1)
hxgy = xlog1x(Pxy @ Py_inv).sum(axis=0) @ py

# return the sum of these
return float(hygx + hxgy)
```

Python 3.5 Pro Tip!

The @ symbols in the above paragraph represent the *matrix multiplication* operator, and were introduced in Python 3.5 in 2015. This is one of the most compelling arguments to use Python 3 for scientific programmers: it enables the programming of linear algebra algorithms using code that remains very close to the original mathematics. Compare the above:

```
hygx = px @ xlog1x(Px_inv @ Pxy).sum(axis=1)
```

with the equivalent Python 2 code:

```
hygx = px.dot(xlog1x(Px_inv.dot(Pxy)).sum(axis=1))
```

By using the @ operator to stay closer to mathematical notation, we can avoid implementation errors and produce code that is much easier to read.

Actually, SciPy's authors knew this long before the @ operator was introduced, and actually altered the meaning of the * operator when the inputs are SciPy matrices. Available in Python 2.7, it lets us produce nice, readable code like the above:

```
hygx = -px * xlog(Px_inv * Pxy).sum(axis=1)
```

But there is a huge catch: this code will behave differently when px or Px_inv are SciPy matrices than when they are not! If Px_inv and Pxy are NumPy arrays, * produces the element-wise multiplication, while if they are SciPy matrices, it produces the matrix product! As you can imagine, this is the source of a great many errors, and much of the SciPy community has abandoned this use in favor of the uglier but unambiguous .dot method.

Python 3.5's @ operator gives us the best of both worlds!

Contingency Tables

But let's start simple and work our way up to segmentations.

Suppose you just started working as a data scientist at email startup Spam-o-matic. You are tasked with building a detector for spam email. You encode the detector outcome as a numeric value, 0 for not spam and 1 for spam.

If you have a set of 10 emails to classify, you end up with a vector of *predictions*:

```
import numpy as np
pred = np.array([0, 1, 0, 0, 1, 1, 1, 0, 1, 1])
```

You can check how well you've done by comparing it to a vector of *ground truth*, classifications obtained by inspecting each message by hand.

```
gt = np.array([0, 0, 0, 0, 0, 1, 1, 1, 1, 1])
```

Now, classification is hard for computers, so the values in pred and gt don't match up exactly. At positions where pred is 0 and gt is 0, the prediction has correctly identified a message as nonspam. This is called a *true negative*. Conversely, at positions where both values are 1, the predictor has correctly identified a spam message and found a *true positive*.

Then, there are two kinds of errors. If we let a spam message (where gt is 1) through to the user's inbox (pred is 0), we've made a *false negative* error. If we predict a legitimate message (gt is 0) to be spam (pred is 1), we've made a *false positive* prediction. (An email from the director of my scientific institute once landed in my spam folder. The reason? His announcement of a postdoc talk competition started with "You could win $500!")

If we want to measure how well we are doing, we have to count the above kinds of errors using a *contingency matrix*. (This is also sometimes called a confusion matrix. The name is apt.) For this, we place the prediction labels along the rows and the ground truth labels along the columns. Then we count the number of times they correspond. So, for example, since there are 4 true positives (where pred and gt are both 1), the matrix will have a value of 3 at position (1, 1).

Generally:

$$C_{i,j} = \Sigma_k \mathbb{I}(p_k = i)\mathbb{I}(g_k = j)$$

Here's an intuitive but inefficient way of building the preceding equation:

```
def confusion_matrix(pred, gt):
    cont = np.zeros((2, 2))
    for i in [0, 1]:
        for j in [0, 1]:
```

```
        cont[i, j] = np.sum((pred == i) & (gt == j))
    return cont
```

We can check that this gives us the right counts:

```
confusion_matrix(pred, gt)
```

```
array([[ 3.,  1.],
       [ 2.,  4.]])
```

Exercise: Computational Complexity of Confusion Matrices

Why did we call this code inefficient?

See "Solution: Computational Complexity of Confusion Matrices" on page 231.

Exercise: Alternative Algorithm to Compute the Confusion Matrix

Write an alternative way of computing the confusion matrix that only makes a single pass through pred and gt.

```
def confusion_matrix1(pred, gt):
    cont = np.zeros((2, 2))
    # your code goes here
    return cont
```

Check out "Solution: Alternative Confusion Matrix Computing" on page 231.

We can make this example a bit more general. Instead of classifying spam and non-spam, we can classify spam, newsletters, sales and promotions, mailing lists, and personal email. That's 5 categories, which we'll label 0 to 4. The confusion matrix will now be 5-by-5, with matches counted on the diagonal, and errors counted on the off-diagonal entries.

The definition of the confusion_matrix function just shown doesn't extend well to this larger matrix, because now we must have *25* passes though the result and ground truth arrays. This problem only grows as we add more email categories, such as social media notifications.

Exercise: Multiclass Confusion Matrix

Write a function to compute the confusion matrix in one pass, as before, but instead of assuming two categories, infer the number of categories from the input.

```
def general_confusion_matrix(pred, gt):
    n_classes = None  # replace `None` with something useful
    # your code goes here
    return cont
```

Your one-pass solution will scale well with the number of classes, but, because the for-loop runs in the Python interpreter, it will be slow when you have a large number of documents. Also, because some classes are easier to mistake for one another, the matrix will be *sparse*, with many 0 entries. Indeed, as the number of classes increases, dedicating lots of memory space to the 0 entries of the contingency matrix is increasingly wasteful. Instead, we can use the `sparse` module of SciPy, which contains objects to efficiently represent sparse matrices.

scipy.sparse Data Formats

We covered the internal data format of NumPy arrays in Chapter 1. We hope you agree that it's a fairly intuitive, and, in some sense, inevitable format to hold n-dimensional array data. For sparse matrices, there are actually a wide array of possible formats, and the "right" format depends on the problem you want to solve. We'll cover the two most commonly used formats, but for a complete list, see the comparison table later in the chapter, as well as the online documentation for `scipy.sparse`.

COO Format

Perhaps the most intuitive is the coordinate, or COO, format. This uses three 1D arrays as follows to represent a 2D matrix A. Each of these arrays has length equal to the number of nonzero values in A, and together they list (i, j, value) coordinates of every entry that is not equal to 0.

- The `row` and `col` arrays, which together specify the location of each nonzero entry (row and column indices, respectively).
- The `data` array, which specifies the *value* at each of those locations.

Every part of the matrix that is not represented by the (`row`, `col`) pairs is considered to be 0. Much more efficient! So, to represent the matrix:

```
s = np.array([[ 4,  0, 3],
              [ 0, 32, 0]], dtype=float)
```

We can do the following:

```
from scipy import sparse

data = np.array([4, 3, 32], dtype=float)
row = np.array([0, 0, 1])
col = np.array([0, 2, 1])

s_coo = sparse.coo_matrix((data, (row, col)))
```

The `.toarray()` method of every sparse format in `scipy.sparse` returns a NumPy array representation of the sparse data. We can use this to check that we created `s_coo` correctly:

```
s_coo.toarray()
```

```
array([[  4.,   0.,   3.],
       [  0.,  32.,   0.]])
```

Identically, we can use the `.A` *property*, which is just like an attribute, but actually executes a function. `.A` is a particularly dangerous property, because it can hide a potentially very large operation: the dense version of a sparse matrix can be orders of magnitude bigger than the sparse matrix itself, bringing a computer to its knees, in just three keystrokes!

```
s_coo.A
```

```
array([[  4.,   0.,   3.],
       [  0.,  32.,   0.]])
```

In this chapter, and elsewhere, we recommend using the `toarray()` method wherever it does not impair readability, as it more clearly signals a potentially expensive operation. However, we will use `.A` where it makes the code more readable by virtue of its brevity (e.g., when trying to implement a sequence of mathematical equations).

Exercise: COO Representation

Write out the COO representation of the following matrix:

```
s2 = np.array([[0, 0, 6, 0, 0],
               [1, 2, 0, 4, 5],
               [0, 1, 0, 0, 0],
               [9, 0, 0, 0, 0],
               [0, 0, 0, 6, 7]])
```

Unfortunately, although the COO format is intuitive, it's not very optimized to use the minimum amount of memory, or to traverse the array as quickly as possible during computations. (Remember from Chapter 1, *data locality* is very important to efficient computation!) However, you can look at your COO representation above to help you identify redundant information. Notice all those repeated 1s?

Compressed Sparse Row Format

If we use COO to enumerate the nonzero entries row-by-row, rather than in arbitrary order (which the format allows), we end up with many consecutive, repeated values in the `row` array. We can compress these by indicating the *indices* in `col` where the next row starts, rather than repeatedly writing the row index. This is the basis for the *compressed sparse row*, or *CSR* format.

Let's work through the example above. In CSR format, the `col` and `data` arrays are unchanged (but `col` is renamed to `indices`). However, the `row` array, instead of indicating the rows, indicates *where* in `col` each row begins, and is renamed to `indptr`, for "index pointer."

So, let's look at `row` and `col` in COO format, ignoring `data`:

```
row = [0, 1, 1, 1, 1, 2, 3, 4, 4]
col = [2, 0, 1, 3, 4, 1, 0, 3, 4]
```

Each new row begins at the index where row changes. The 0th row starts at index 0, and the 1st row starts at index 1, but the 2nd row starts where "2" first appears in row, at index 5. Then, the indices increase by 1 for rows 3 and 4, to 6 and 7. The final index, indicating the end of the matrix, is the total number of nonzero values (9). So:

```
indptr = [0, 1, 5, 6, 7, 9]
```

Let's use these hand-computed arrays to build a CSR matrix in SciPy. We can check our work by comparing the .A output from our COO and CSR representations to the NumPy array s2 that we defined earlier.

```
data = np.array([6, 1, 2, 4, 5, 1, 9, 6, 7])

coo = sparse.coo_matrix((data, (row, col)))
csr = sparse.csr_matrix((data, col, indptr))

print('The COO and CSR arrays are equal: ',
      np.all(coo.A == csr.A))
print('The CSR and NumPy arrays are equal: ',
      np.all(s2 == csr.A))

The COO and CSR arrays are equal:  True
The CSR and NumPy arrays are equal:  True
```

The ability to store large, sparse matrices, and perform computations on them, is incredibly powerful and can be applied in many domains.

For example, one can think of the entire web as a large, sparse, $N \times N$ matrix. Each entry X_{ij} indicates whether web page i links to page j. By normalizing this matrix and solving for its dominant eigenvector, one obtains the so-called PageRank—one of the numbers Google uses to order your search results. (You can read more about this in the next chapter.)

As another example, we can represent the human brain as a large $m \times m$ graph, where there are m nodes (positions) in which you measure activity using an MRI scanner. After a while of measuring, correlations can be calculated and entered into a matrix C_{ij}. Thresholding this matrix produces a sparse matrix of ones and zeros. The eigenvector corresponding to the second-smallest eigenvalue of this matrix partitions the m brain areas into subgroups, which, it turns out, are often related to functional regions of the brain![1]

---

1 M. E. J. Newman, "Modularity and Community Structure in Networks" (*http://dx.doi.org/DOI:10.1073/pnas. 0601602103*), *PNAS* 103, no. 23 (2006):8577–8582.

	bsr_matrix	coo_matrix	csc_matrix	csr_matrix	dia_matrix	dok_matrix	lil_matrix
Full name	Block Sparse Row	Coordinate	Compressed Sparse Column	Compressed Sparse Row	Diagonal	Dictionary of Keys	Row-based linked-list
Note	Similar to CSR	Only used to construct sparse matrices, which are then converted to CSC or CSR for further operations.				Used to construct sparse matrices incrementally	Used to construct sparse matrices incrementally
Use cases	• Storage of dense submatrices • Often used in numerical analyses of discretized problems, such as finite elements, differential equations	• Fast and straightforward way of constructing sparse matrices • During construction, duplicate coordinates are summed—useful for, e.g., finite element analysis	• Arithmetic operations (supports addition, subtraction, multiplication, division, and matrix power) • Efficient column slicing • Fast matrix-vector products (CSR, BSR can be faster, depending on the problem)	• Arithmetic operations • Efficient row slicing • Fast matrix-vector products	• Arithmetic operations	• Changes in sparsity structure are inexpensive • Arithmetic operations • Fast access to individual elements • Efficient conversion to COO (but no duplicates allowed)	• Changes in sparsity structure are inexpensive • Flexible slicing
Cons		• No arithmetic operations • No slicing	• Slow row slicing (see CSR) • Changes to sparsity structure are expensive (see LIL, DOK)	• Slow column slicing (see CSC) • Changes to sparsity structure are expensive (see LIL, DOK)	• Sparsity structure limited to values on diagonals	• Expensive for arithmetic operations • Slow matrix-vector products	• Expensive for arithmetic operations • Slow column slicing • Slow matrix-vector products

Applications of Sparse Matrices: Image Transformations

Libraries like scikit-image and SciPy already contain algorithms for transforming (rotating and warping) images effectively, but what if you were head of the NumPy Agency for Space Affairs and had to rotate millions of images streaming in from the newly launched Jupyter Orbiter?

In such cases, you want to squeeze every ounce of performance from your computer. It turns out that we can do a lot better than even the optimized C code in SciPy's ndimage if we are repeatedly applying the *same* transformation.

We'll use the following test image of a cameraman from scikit-image as example data:

```
# Make plots appear inline, set custom plotting style
%matplotlib inline
import matplotlib.pyplot as plt
plt.style.use('style/elegant.mplstyle')

from skimage import data
image = data.camera()
plt.imshow(image);
```

As a test operation, we'll be rotating the image by 30 degrees. We begin by defining the transformation matrix, H, which, when multiplied with a coordinate from the input image, $[r, c, 1]$, will give us the corresponding coordinate in the output, $[r', c', 1]$. (Note: we are using homogeneous coordinates (*https://en.wikipedia.org/wiki/Homogeneous_coordinates*), which have a 1 appended to them and which give us greater flexibility when we're defining linear transforms.)

```
angle = 30
c = np.cos(np.deg2rad(angle))
s = np.sin(np.deg2rad(angle))

H = np.array([[c, -s,  0],
              [s,  c,  0],
              [0,  0,  1]])
```

You can verify that this works by multiplying H with the point (1, 0). A 30-degree counterclockwise rotation around the origin (0, 0) should take us to point $(\frac{\sqrt{3}}{2}, \frac{1}{2})$:

```
point = np.array([1, 0, 1])
print(np.sqrt(3) / 2)
print(H @ point)
```

```
0.866025403784
[ 0.8660254  0.5        1.        ]
```

Similarly, applying the 30-degree rotation three times should get us to the column axis, at point (0, 1). We can see that this works, minus some floating-point approximation error:

```
print(H @ H @ H @ point)
```

```
[ 2.77555756e-16   1.00000000e+00   1.00000000e+00]
```

Now, we will build a function that defines a "sparse operator." The goal of the sparse operator is to take all pixels of the output image, figure out where they came from in the input image, and do the appropriate (bilinear) interpolation (see Figure 5-1) to calculate their values. It does this using just matrix multiplication on the image values, and thus is extremely fast.

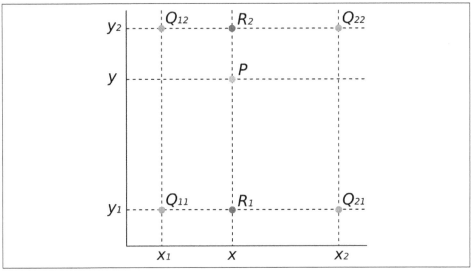

Figure 5-1. Diagram explaining bilinear interpolation—the value at point P is estimated as a weighted sum of the values at Q_{11}, Q_{12}, Q_{21}, Q_{22}

Let's look at the function that builds our sparse operator:

```
from itertools import product

def homography(tf, image_shape):
```

```
"""Represent homographic transformation & interpolation as linear operator.

Parameters
----------
tf : (3, 3) ndarray
    Transformation matrix.
image_shape : (M, N)
    Shape of input gray image.

Returns
-------
A : (M * N, M * N) sparse matrix
    Linear-operator representing transformation + bilinear interpolation.

"""
# Invert matrix.  This tells us, for each output pixel, where to
# find its corresponding input pixel.
H = np.linalg.inv(tf)

m, n = image_shape

# We are going to construct a COO matrix, often called IJK matrix,
# for which we'll need row coordinates (I), column coordinates (J),
# and values (K).
row, col, values = [], [], []

# For each pixel in the output image...
for sparse_op_row, (out_row, out_col) in \
        enumerate(product(range(m), range(n))):

    # Compute where it came from in the input image
    in_row, in_col, in_abs = H @ [out_row, out_col, 1]
    in_row /= in_abs
    in_col /= in_abs

    # if the coordinates are outside of the original image, ignore this
    # coordinate; we will have 0 at this position
    if (not 0 <= in_row < m - 1 or
            not 0 <= in_col < n - 1):
        continue

    # We want to find the four surrounding pixels, so that we
    # can interpolate their values to find an accurate
    # estimation of the output pixel value.
    # We start with the top, left corner, noting that the remaining
    # points are 1 away in each direction.
    top = int(np.floor(in_row))
    left = int(np.floor(in_col))

    # Calculate the position of the output pixel, mapped into
    # the input image, within the four selected pixels.
    # https://commons.wikimedia.org/wiki/File:BilinearInterpolation.svg
```

```
        t = in_row - top
        u = in_col - left

        # The current row of the sparse operator matrix is given by the
        # raveled output pixel coordinates, contained in sparse_op_row.
        # We will take the weighted average of the four surrounding input
        # pixels, corresponding to four columns. So we need to repeat the row
        # index four times.
        row.extend([sparse_op_row] * 4)

        # The actual weights are calculated according to the bilinear
        # interpolation algorithm, as shown at
        # https://en.wikipedia.org/wiki/Bilinear_interpolation
        sparse_op_col = np.ravel_multi_index(
                ([top, top,     top + 1, top + 1 ],
                 [left, left + 1, left,    left + 1]), dims=(m, n))
        col.extend(sparse_op_col)
        values.extend([(1-t) * (1-u), (1-t) * u, t * (1-u), t * u])

    operator = sparse.coo_matrix((values, (row, col)),
                                 shape=(m*n, m*n)).tocsr()
    return operator
```

Recall that we apply the sparse operator as follows:

```
def apply_transform(image, tf):
    return (tf @ image.flat).reshape(image.shape)
```

Let's try it out!

```
tf = homography(H, image.shape)
out = apply_transform(image, tf)
plt.imshow(out);
```

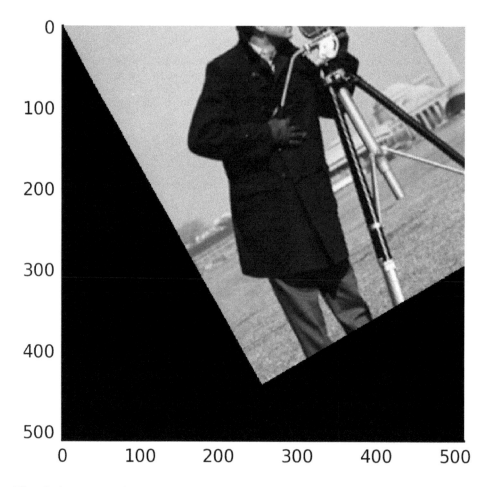

There's that rotation!

Exercise: Image Rotation

The rotation happens around the origin, coordinates (0, 0). But can you rotate the image around its center?

Hint: The transformation matrix for a *translation* (i.e., sliding the image up/down or left/right), is given by:

$$H_{tr} = \begin{bmatrix} 1 & 0 & t_r \\ 0 & 1 & t_c \\ 0 & 0 & 1 \end{bmatrix}$$

when you want to move the image t_r pixels down and t_c pixels right.

As previously mentioned, this sparse linear operator approach to image transformation is fast. Let's measure how it performs in comparison to ndimage. To make the comparison fair, we need to tell ndimage that we want linear interpolation with order=1, and that we want to ignore pixels outside of the original shape, with reshape=False.

```
%timeit apply_transform(image, tf)

100 loops, average of 7: 3.35 ms +- 270 µs per loop (using standard deviation)

from scipy import ndimage as ndi
%timeit ndi.rotate(image, 30, reshape=False, order=1)

100 loops, average of 7: 19.7 ms +- 988 µs per loop (using standard deviation)
```

On our machines, we see a speed-up of approximately 10 times. While this example does only a rotation, there is no reason why we cannot do more complicated warping operations, such as correcting for a distorted lens during imaging, or making people pull funny faces. Once the transform has been computed, applying it repeatedly is fast, thanks to sparse matrix algebra.

So now that we've seen a "standard" use of SciPy's sparse matrices, let's have a look at the out-of-the-box use that inspired this chapter.

Back to Contingency Tables

You might recall that we are trying to quickly build a sparse, joint probability matrix using SciPy's sparse formats. We know that the COO format stores sparse data as three arrays, containing the row and column coordinates of nonzero entries, as well as their values. But we can use a little known feature of COO to obtain our matrix extremely quickly.

Have a look at this data:

```
row = [0, 0, 2]
col = [1, 1, 2]
dat = [5, 7, 1]
S = sparse.coo_matrix((dat, (row, col)))
```

Notice that the entry at (row, column) position (0, 1) appears twice: first as 5 and then at 7. What should the matrix value at (0, 1) be? Cases could be made for both the earliest entry encountered, or the latest, but what was in fact chosen is the *sum*:

```
print(S.toarray())

[[ 0 12  0]
 [ 0  0  0]
 [ 0  0  1]]
```

So, COO format will sum together repeated entries. Which is exactly what we need to do to make a contingency matrix! Indeed, our task is pretty much done: we can set pred as the rows, gt as the columns, and simply 1 as the values. The ones will get summed together and count the number of times that label *i* in pred occurs together with label *j* in gt at position *i, j* in the matrix! Let's try it out:

```
from scipy import sparse

def confusion_matrix(pred, gt):
    cont = sparse.coo_matrix((np.ones(pred.size), (pred, gt)))
    return cont
```

To look at a small one, we simply use the .toarray method, as above:

```
cont = confusion_matrix(pred, gt)
print(cont)

  (0, 0)    1.0
  (1, 0)    1.0
  (0, 0)    1.0
  (0, 0)    1.0
  (1, 0)    1.0
  (1, 1)    1.0
  (1, 1)    1.0
  (0, 1)    1.0
  (1, 1)    1.0
  (1, 1)    1.0

print(cont.toarray())

[[ 3.  1.]
 [ 2.  4.]]
```

It works!

Exercise: Reducing the Memory Footprint

Remember from Chapter 1 that NumPy has built-in tools for repeating arrays using *broadcasting*. How can you reduce the memory footprint required for the contingency matrix computation?

Hint: Look at the documentation for the function np.broadcast_to.

Contingency Tables in Segmentation

You can think of the segmentation of an image in the same way as the classification problem above: the segment label at each *pixel* is a *prediction* about which *class* the pixel belongs to. And NumPy arrays allow us to do this transparently, because their .ravel() method returns a 1D view of the underlying data.

As an example, here's a segmentation of a tiny 3 × 3 image:

```
seg = np.array([[1, 1, 2],
                [1, 2, 2],
                [3, 3, 3]], dtype=int)
```

Here's the ground truth, what some person said was the correct way to segment this image:

```
gt = np.array([[1, 1, 1],
               [1, 1, 1],
               [2, 2, 2]], dtype=int)
```

We can think of these two as classifications, just like before. Every pixel is a different prediction.

```
print(seg.ravel())
print(gt.ravel())

[1 1 2 1 2 2 3 3 3]
[1 1 1 1 1 1 2 2 2]
```

Then, like above, the contingency matrix is given by:

```
cont = sparse.coo_matrix((np.ones(seg.size),
                          (seg.ravel(), gt.ravel())))
print(cont)

  (1, 1)    1.0
  (1, 1)    1.0
  (2, 1)    1.0
  (1, 1)    1.0
  (2, 1)    1.0
  (2, 1)    1.0
  (3, 2)    1.0
  (3, 2)    1.0
  (3, 2)    1.0
```

Some indices appear more than once, but we can use the summing feature of the COO format to confirm that this represents the matrix we want:

```
print(cont.toarray())

[[ 0.  0.  0.]
 [ 0.  3.  0.]
 [ 0.  3.  0.]
 [ 0.  0.  3.]]
```

How do we convert this table into a measure of how well seg represents gt? Segmentation is a hard problem, so it's important to measure how well a segmentation algorithm is doing, by comparing its output to a "ground truth" segmentation that is manually produced by a human.

But even this comparison is not an easy task. How do we define how "close" an automated segmentation is to a ground truth? We'll illustrate one method, the *variation of information*, or VI (Meila, 2005). This is defined as the answer to the following question: on average, for a random pixel, if we are given its segment ID in one segmentation, how much more *information* do we need to determine its ID in the other segmentation?

Intuitively, if the two segmentations are exactly alike, then knowing the segment ID in one tells you the segment ID in the other, with no additional information. But as the segmentations become more different, knowing an ID in one doesn't tell you the ID in the other without more information.

Information Theory in Brief

In order to answer this question, we'll need a quick primer on information theory. We need to be brief but if you want more information (heh), you should look at Christopher Olah's stellar blog post, Visual Information Theory (*https://colah.github.io/posts/2015-09-Visual-Information/*).

The basic unit of information is the *bit*, commonly shown as a 0 or 1, representing an equal probability choice between two options. This is straightforward: if I want to tell you whether a coin toss landed as heads or tails, I need one bit, which can take many forms: a long or short pulse over a telegraph wire (as in Morse code), a light flashing one of two colors, or a single number taking values 0 or 1. Importantly, I *always* need one bit, because the outcome of a coin toss is random.

It turns out that we can extend this concept to *fractional* bits for events that are *less* random. Suppose, for example, that you need to transmit whether it rained today in Los Angeles. At first glance, it seems that this requires 1 bit as well: 0 for it didn't rain, 1 for it rained. However, rain in LA is a rare event, so over time we can actually get away with transmitting much less information: transmit a 0 *occasionally* just to make sure that our communication is still working, but otherwise simply *assume* that the signal is 0, and send 1 only on those rare occasions that it rains.

Thus, when two events are *not* equally likely, we need *less* than 1 bit to represent them. Generally, we measure this for any random variable X (which could have more than two possible values) by using the *entropy* function H:

$$H(X) = \sum_x p_x \log_2 \left(\frac{1}{p_x} \right)$$
$$= -\sum_x p_x \log_2 \left(p_x \right)$$

where the xs are possible values of X, and p_x is the probability of X taking value x.

So, the entropy of a coin toss T that can take values heads (h) and tails (t) is:

$$H(T) = p_h log_2(1/p_h) + p_t log_2(1/p_t)$$
$$= 1/2 log_2(2) + 1/2 log_2(2)$$
$$= 1/2 \cdot 1 + 1/2 \cdot 1$$
$$= 1$$

$$H(T) = p_h log_2(1/p_h) + p_t log_2(1/p_t)$$
$$= 1/2 log_2(2) + 1/2 log_2(2)$$
$$= 1/2 \cdot 1 + 1/2 \cdot 1$$
$$= 1$$

The long-term probability of rain on any given day in LA is about 1 in 6, so the entropy of rain in LA, R, taking values rain (r) or shine (s) is:

$$H(R) = p_r log_2(1/p_r) + p_s log_2(1/p_s)$$
$$= 1/6 log_2(6) + 5/6 log_2(6/5)$$
$$\approx 0.65 \text{ bits}$$

A special kind of entropy is the *conditional* entropy. This is the entropy of a variable *assuming* that you also know something else about that variable. For example, what is the entropy of rain *given* that you know the month? This is written as:

$$H(R|M) = \Sigma_{m=1}^{12} p(m)H(R|M = m)$$

and:

$$H(R|M = m) = p_{r|m} log_2 \left(\frac{1}{p_{r|m}} \right) + p_{s|m} log_2 \left(\frac{1}{p_{s|m}} \right)$$
$$= \frac{p_{rm}}{p_m} log_2 \left(\frac{p_m}{p_{rm}} \right) + \frac{p_{sm}}{p_m} log_2 \left(\frac{p_m}{p_{sm}} \right)$$
$$= -\frac{p_{rm}}{p_m} log_2 \left(\frac{p_{rm}}{p_m} \right) - \frac{p_{sm}}{p_m} log_2 \left(\frac{p_{sm}}{p_m} \right)$$

You now have all the information theory you need to understand the variation of information. In the preceding example, events are days, and they have two properties:

- rain/shine
- month

By observing many days, we can build a *contingency matrix*, just like the ones in the classification examples, measuring the month of a day and whether it rained. We're not going to travel to LA to do this (fun as it would be), and instead we use the historical table that follows, roughly eyeballed from WeatherSpark (*http://bit.ly/2sXj4D9*):

Month	P(rain)	P(shine)
1	0.25	0.75
2	0.27	0.73
3	0.24	0.76
4	0.18	0.82
5	0.14	0.86
6	0.11	0.89
7	0.07	0.93
8	0.08	0.92
9	0.10	0.90
10	0.15	0.85
11	0.18	0.82
12	0.23	0.77

The conditional entropy of `rain` given `month` is then:

$$H(R|M) = -\frac{1}{12}\left(0.25\log_2(0.25) + 0.75\log_2(0.75)\right) - \frac{1}{12}\left(0.27\log_2(0.27) + 0.73\log_2(0.73)\right)$$

$$-\dots - \frac{1}{12}\left(0.23\log_2(0.23) + 0.77\log_2(0.77)\right)$$

$$\approx 0.626 \text{ bits}$$

So, by using the month, we've reduced the randomness of the signal, but not by much!

We can also compute the conditional entropy of `month` given `rain`, which measures how much information we need to determine the month if we know it rained. Intuitively, we know that this is better than going in blind, since it's more likely to rain in the winter months.

Exercise: Computing Conditional Entropy

Compute the conditional entropy of month given rain. What is the entropy of the month variable? (Ignore the different number of days in a month.) Which one is greater?

The probabilities in the table are the conditional probabilities of rain given month.)

```
prains = np.array([25, 27, 24, 18, 14, 11, 7, 8, 10, 15, 18, 23]) / 100
pshine = 1 - prains
p_rain_g_month = np.column_stack([prains, pshine])
# replace 'None' below with expression for nonconditional contingency
# table. Hint: the values in the table must sum to 1.
p_rain_month = None
# Add your code below to compute H(M|R) and H(M)
```

Together, these two values define the variation of information (VI):

$$VI(A, B) = H(A|B) + H(B|A)$$

Information Theory in Segmentation: Variation of Information

Back in the image segmentation context, "days" become "pixels," and "rain" and "month" become "label in automated segmentation (S)" and "label ground truth (T)." Then, the conditional entropy of the automatic segmentation given the ground truth measures how much additional information we need to determine a pixel's identity in S if we are told its identity in T. For example, if every T segment g is split into two equally sized segments a_1 and a_2 in S, then $H(S|T) = 1$, because after knowing a pixel is in g, you still need 1 additional bit to know whether it belongs to a_1 or a_2. However, $H(T|S) = 0$, because regardless of whether a pixel is in a_1 or a_2, it is guaranteed to be in g, so you need no more information than the segment in S.

So, together, in this case:

$$VI(S, T) = H(S|T) + H(T|S) = 1 + 0 = 1 \text{ bit}$$

Here's a simple example:

```
S = np.array([[0, 1],
              [2, 3]], int)

T = np.array([[0, 1],
              [0, 1]], int)
```

Here we have two segmentations of a four-pixel image: S and T. S puts every pixel in its own segment, while T puts the left two pixels in segment 0 and the right two pixels in segment 1. Now, we make a contingency table of the pixel labels, just as we did with the spam-prediction labels. The only difference is that the label arrays are 2D, instead of the 1D arrays of predictions. In fact, this doesn't matter: remember that NumPy arrays are actually linear (1D) chunks of data with some shape and other metadata attached. As we mentioned before, we can ignore the shape by using the arrays' .ravel() method:

```
S.ravel()

array([0, 1, 2, 3])
```

Now we can just make the contingency table in the same way as when we were predicting spam:

```
cont = sparse.coo_matrix((np.broadcast_to(1., S.size),
                          (S.ravel(), T.ravel())))
cont = cont.toarray()
cont

array([[ 1.,  0.],
       [ 0.,  1.],
       [ 1.,  0.],
       [ 0.,  1.]])
```

In order to make this a table of probabilities, instead of counts, we simply divide by the total number of pixels:

```
cont /= np.sum(cont)
```

Finally, we can use this table to compute the probabilities of labels in *either* S or T using the axis-wise sums:

```
p_S = np.sum(cont, axis=1)
p_T = np.sum(cont, axis=0)
```

There is a small kink in writing Python code to compute entropy: although 0 log(0) is defined to be equal to 0, in Python, it is undefined, and results in a nan (not a number) value:

```
print('The log of 0 is: ', np.log2(0))
print('0 times the log of 0 is: ', 0 * np.log2(0))

The log of 0 is:  -inf
0 times the log of 0 is:  nan
```

Therefore, we have to use NumPy indexing to mask out the 0 values. Additionally, we'll need a slightly different strategy depending on whether the input is a NumPy array or a SciPy sparse matrix. We'll write the following convenience function:

```
def xlog1x(arr_or_mat):
    """Compute the element-wise entropy function of an array or matrix.

    Parameters
    ----------
    arr_or_mat : numpy array or scipy sparse matrix
```

```
        The input array of probabilities. Only sparse matrix formats with a
        `data` attribute are supported.

    Returns
    -------
    out : array or sparse matrix, same type as input
        The resulting array. Zero entries in the input remain as zero,
        all other entries are multiplied by the log (base 2) of their
        inverse.
    """
    out = arr_or_mat.copy()
    if isinstance(out, sparse.spmatrix):
        arr = out.data
    else:
        arr = out
    nz = np.nonzero(arr)
    arr[nz] *= -np.log2(arr[nz])
    return out
```

Let's make sure it works:

```
a = np.array([0.25, 0.25, 0, 0.25, 0.25])
xlog1x(a)

array([ 0.5,  0.5,  0. ,  0.5,  0.5])

mat = sparse.csr_matrix([[0.125, 0.125, 0.25,    0],
                         [0.125, 0.125,    0, 0.25]])
xlog1x(mat).A

array([[ 0.375,  0.375,  0.5 ,  0.  ],
       [ 0.375,  0.375,  0.  ,  0.5 ]])
```

So, the conditional entropy of *S* given *T*:

```
H_ST = np.sum(np.sum(xlog1x(cont / p_T), axis=0) * p_T)
H_ST

1.0
```

And the converse:

```
H_TS = np.sum(np.sum(xlog1x(cont / p_S[:, np.newaxis]), axis=1) * p_S)
H_TS

0.0
```

Converting NumPy Array Code to Use Sparse Matrices

We used NumPy arrays and broadcasting in the above examples, which, as we've seen many times, is a powerful way to analyze data in Python. However, for segmentations of complex images, possibly containing thousands of segments, it rapidly becomes inefficient. We can instead use sparse throughout the calculation, and recast some of the NumPy magic as linear algebra operations. This was suggested (*http://bit.ly/2trePTS*) to us by Warren Weckesser on StackOverflow.

The linear algebra version efficiently computes a contingency matrix for very large amounts of data, up to billions of points, and is elegantly concise.

```python
import numpy as np
from scipy import sparse


def invert_nonzero(arr):
    arr_inv = arr.copy()
    nz = np.nonzero(arr)
    arr_inv[nz] = 1 / arr[nz]
    return arr_inv


def variation_of_information(x, y):
    # compute contingency matrix, aka joint probability matrix
    n = x.size
    Pxy = sparse.coo_matrix((np.full(n, 1/n), (x.ravel(), y.ravel())),
                            dtype=float).tocsr()

    # compute marginal probabilities, converting to 1D array
    px = np.ravel(Pxy.sum(axis=1))
    py = np.ravel(Pxy.sum(axis=0))

    # use sparse matrix linear algebra to compute VI
    # first, compute the inverse diagonal matrices
    Px_inv = sparse.diags(invert_nonzero(px))
    Py_inv = sparse.diags(invert_nonzero(py))

    # then, compute the entropies
    hygx = px @ xlog1x(Px_inv @ Pxy).sum(axis=1)
    hxgy = xlog1x(Pxy @ Py_inv).sum(axis=0) @ py

    # return the sum of these
    return float(hygx + hxgy)
```

We can check that this gives the right value (1) for the VI of our toy S and T:

```python
variation_of_information(S, T)
```

```
1.0
```

You can see how we use three types of sparse matrices (COO, CSR, and diagonal) to efficiently solve the entropy calculation in the case of sparse contingency matrices, where NumPy would be inefficient. (Indeed, this whole approach was inspired by a Python MemoryError!)

Using Variation of Information

To finish, let's demonstrate the use of VI to estimate the best possible automated segmentation of an image. You may remember our friendly stalking tiger from Chapter 3 (see Figure 5-2). (If you don't, you might want to work on your threat-assessment skills!) Using our skills from Chapter 3, we're going to generate a number of possible ways of segmenting the tiger image, and then figure out the best one.

```
from skimage import io

url = ('http://www.eecs.berkeley.edu/Research/Projects/CS/vision/bsds'
       '/BSDS300/html/images/plain/normal/color/108073.jpg')
tiger = io.imread(url)

plt.imshow(tiger);
```

Figure 5-2. BSDS tiger image, number 108073

In order to check our image segmentation, we're going to need some ground truth. It turns out that humans are awesome at detecting tigers (natural selection for the win!), so all we need to do is ask a human to find the tiger. Luckily, researchers at Berkeley have already asked dozens of humans to look at this image and manually segment it.[2]

---

2 Pablo Arbelaez, Michael Maire, Charless Fowlkes, and Jitendra Malik, "Contour Detection and Hierarchical Image Segmentation," *IEEE TPAMI* 33, no. 5 (2011): 898–916.

Let's grab one of the segmentation images from the Berkeley Segmentation Dataset and Benchmark (*http://bit.ly/2sdHN92*) (see Figure 5-3). It's worth noting that there is quite substantial variation between the segmentations performed by humans. If you look through the various tiger segmentations (*http://bit.ly/2sdWtoH*), you will see that some humans are more pedantic than others about tracing around the reeds, while others consider the reflections to be objects worth segmenting out from the rest of the water. We have chosen a segmentation that we like (one with pedantic reed tracing, because we are perfectionistic scientist-types). But to be clear, we really have no single ground truth!

```
from scipy import ndimage as ndi
from skimage import color

human_seg_url = ('http://www.eecs.berkeley.edu/Research/Projects/CS/'
                 'vision/bsds/BSDS300/html/images/human/normal/'
                 'outline/color/1122/108073.jpg')
boundaries = io.imread(human_seg_url)
plt.imshow(boundaries);
```

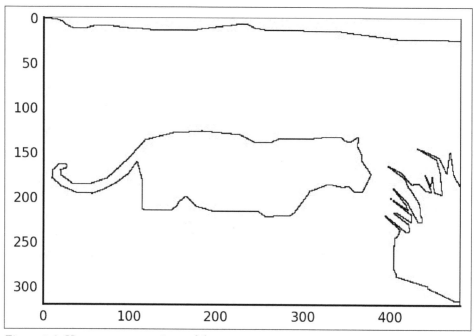

Figure 5-3. Human segmentations of the tiger image

Overlaying the tiger image with the human segmentation, we can see that (unsurprisingly) this person does a pretty good job of finding the tiger (see Figure 5-4). They have also segmented out the river bank and a tuft of reeds. Nice job, human #1122!

```
human_seg = ndi.label(boundaries > 100)[0]
plt.imshow(color.label2rgb(human_seg, tiger));
```

Figure 5-4. Human segmentation of the tiger image, overlaid

Now, let's grab our image segmentation code from Chapter 3, and see how well a
Python does at recognizing a tiger (Figure 5-5)!

```
# Draw a region adjacency graph (RAG) - all code from Ch3
import networkx as nx
import numpy as np
from skimage.future import graph

def add_edge_filter(values, graph):
    current = values[0]
    neighbors = values[1:]
    for neighbor in neighbors:
        graph.add_edge(current, neighbor)
    return 0. # generic_filter requires a return value, which we ignore!

def build_rag(labels, image):
    g = nx.Graph()
    footprint = ndi.generate_binary_structure(labels.ndim, connectivity=1)
    for j in range(labels.ndim):
        fp = np.swapaxes(footprint, j, 0)
        fp[0, ...] = 0  # zero out top of footprint on each axis
    _ = ndi.generic_filter(labels, add_edge_filter, footprint=footprint,
                           mode='nearest', extra_arguments=(g,))
```

```
    for n in g:
        g.node[n]['total color'] = np.zeros(3, np.double)
        g.node[n]['pixel count'] = 0
    for index in np.ndindex(labels.shape):
        n = labels[index]
        g.node[n]['total color'] += image[index]
        g.node[n]['pixel count'] += 1
    return g

def threshold_graph(g, t):
    to_remove = ((u, v) for (u, v, d) in g.edges(data=True)
                 if d['weight'] > t)
    g.remove_edges_from(to_remove)

# Baseline segmentation
from skimage import segmentation
seg = segmentation.slic(tiger, n_segments=30, compactness=40.0,
                        enforce_connectivity=True, sigma=3)
plt.imshow(color.label2rgb(seg, tiger));
```

Figure 5-5. Baseline SLIC segmentation of the tiger image

In Chapter 3, we set the graph threshold at 80 and sort of hand-waved over the whole thing. Now we're going to take a closer look at how this threshold impacts our segmentation accuracy. Let's pop the segmentation code into a function so we can play with it.

```
def rag_segmentation(base_seg, image, threshold=80):
    g = build_rag(base_seg, image)
    for n in g:
        node = g.node[n]
        node['mean'] = node['total color'] / node['pixel count']
    for u, v in g.edges_iter():
        d = g.node[u]['mean'] - g.node[v]['mean']
        g[u][v]['weight'] = np.linalg.norm(d)

    threshold_graph(g, threshold)

    map_array = np.zeros(np.max(seg) + 1, int)
    for i, segment in enumerate(nx.connected_components(g)):
        for initial in segment:
            map_array[int(initial)] = i
    segmented = map_array[seg]
    return(segmented)
```

Let's try a few thresholds and see what happens (see Figures 5-6 and 5-7):

```
auto_seg_10 = rag_segmentation(seg, tiger, threshold=10)
plt.imshow(color.label2rgb(auto_seg_10, tiger));
```

Figure 5-6. Tiger RAG-based segmentation at threshold 10

```
auto_seg_40 = rag_segmentation(seg, tiger, threshold=40)
plt.imshow(color.label2rgb(auto_seg_40, tiger));
```

Figure 5-7. Tiger RAG-based segmentation at threshold 40

Actually, in Chapter 3 we did the segmentation a bunch of times with different thresholds and then (because we're human, so we can) picked one that produced a good segmentation. This is a completely unsatisfying way to program image segmentation. Clearly, we need a way to automate this.

We can see that the higher threshold seems to produce a better segmentation. But we have a ground truth, so we can actually put a number to this! Using all our sparse matrix skills, we can calculate the VI for each segmentation.

```
variation_of_information(auto_seg_10, human_seg)
```

```
3.44884607874861
```

```
variation_of_information(auto_seg_40, human_seg)
```

```
1.0381218706889725
```

The high threshold has a smaller variation of information, so it's a better segmentation! Now we can calculate the VI for a range of possible thresholds and see which one gives us the closest segmentation to the human ground truth (Figure 5-8).

```
# Try many thresholds
def vi_at_threshold(seg, tiger, human_seg, threshold):
    auto_seg = rag_segmentation(seg, tiger, threshold)
    return variation_of_information(auto_seg, human_seg)
```

```
thresholds = range(0, 110, 10)
vi_per_threshold = [vi_at_threshold(seg, tiger, human_seg, threshold)
                    for threshold in thresholds]

plt.plot(thresholds, vi_per_threshold);
```

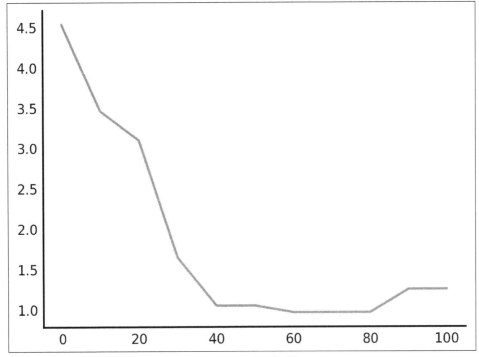

Figure 5-8. Segmentation VI as a function of threshold

Unsurprisingly, it turns out that eyeballing it and picking `threshold=80` did give us one of the best segmentations (Figure 5-9). But now we have a way to automate this process for any image!

```
auto_seg = rag_segmentation(seg, tiger, threshold=80)
plt.imshow(color.label2rgb(auto_seg, tiger));
```

Figure 5-9. Optimal tiger segmentation based on the VI curve

Further Work: Segmentation in Practice

Try finding the best threshold for a selection of other images from the Berkeley Segmentation Dataset and Benchmark (*http://bit.ly/2sdHN92*).[3] Using the mean or median of those thresholds, go and segment a new image. Did you get a reasonable segmentation?

Sparse matrices are an efficient way of representing data with many gaps—a situation that occurs surprisingly often. After reading this chapter, you'll probably start noticing opportunities to use them all the time...and you'll know how.

One particular situation where sparse matrices come extremely handy is in sparse linear algebra. Read on to the next chapter to find out more!

---

3 Pablo Arbelaez, Michael Maire, Charless Fowlkes, and Jitendr Malik, "Contour Detection and Hierarchical Image Segmentation," *IEEE TPAMI*, 33, no. 5, (2011): 898–916.

## CHAPTER 6

Linear Algebra in SciPy

No one can be told what the matrix is. You have to see it for yourself.
   —Morpheus, *The Matrix*

Just like Chapter 4, which dealt with the FFT, this chapter will feature an elegant *method*. We want to highlight the packages available in SciPy to do linear algebra, which forms the basis of much scientific computing.

Linear Algebra Basics

A chapter in a programming book is not really the right place to learn about linear algebra itself, so we assume familiarity with linear algebra concepts. At minimum, you should know that linear algebra involves vectors (ordered collections of numbers) and transforming them by multiplying them with matrices (collections of vectors). If all of this sounds like gibberish to you, you should probably pick up an introductory linear algebra textbook before reading this. We highly recommend Gil Strang's *Linear Algebra and Its Applications* (Pearson, 1994). An introduction is all you need, though—we hope to convey the power of linear algebra while keeping the operations relatively simple!

As an aside, we will break Python notation convention in order to match linear algebra conventions. In Python, variable names should usually begin with a lowercase letter. However, in linear algebra, matrices are denoted by a capital letter, while vectors and scalar values are lowercase. Since we're going to be dealing with quite a few matrices and vectors, following the linear algebra convention helps to keep them straight. Therefore, variables that represent matrices will start with a capital letter, while vectors and numbers will start with lowercase:

```
import numpy as np
```

```
m, n = (5, 6)  # scalars
M = np.ones((m, n))  # a matrix
v = np.random.random((n,))  # a vector
w = M @ v  # another vector
```

In mathematical notation, the vectors would typically be written in boldface, as in **v** and **w**, while the scalars would not, as in *m* and *n*. In Python code, we can't make that distinction, so we will rely instead on context to keep scalars and vectors straight.

Laplacian Matrix of a Graph

We discussed graphs in Chapter 3, where we represented image regions as nodes, connected by edges between them. But we used a rather simple method of analysis: we *thresholded* the graph, removing all edges above some value. Thresholding works in simple cases, but can easily fail, because all you need is one value to fall on the wrong side of the threshold for the approach to fail.

As an example, suppose you are at war, and your enemy is camped just across the river from your forces. You want to cut them off, so you decide to blow up all the bridges between you. Intelligence suggests that you need *t* kg of TNT to blow each bridge crossing the river, but the bridges in your own territory can withstand *t* + 1 kg. You might, having read Chapter 3, order your commandos to detonate *t* kg of TNT on every bridge in the region. But, if intelligence was wrong about just *one* bridge crossing the river, and it remains standing, the enemy's army can come marching through! Disaster!

So, in this chapter, we will explore some alternative approaches to graph analysis, based on linear algebra. It turns out that we can think of a graph, *G*, as an *adjacency matrix*, in which we number the nodes of the graph from 0 to *n* − 1, and place a 1 in row *i*, column *j* of the matrix whenever there is an edge from node *i* to node *j*. In other words, if we call the adjacency matrix *A*, then $A_{i,j} = 1$ if and only if the edge (*i*, *j*) is in *G*. We can then use linear algebra techniques to study this matrix, often with striking results.

The *degree* of a node is defined as the number of edges touching it. For example, if a node is connected to five other nodes in a graph, its degree is 5. (Later, we will differentiate between out-degree and in-degree, when edges have a "from" and "to.") In matrix terms, the degree corresponds to the *sum* of the values in a row or column.

The *Laplacian* matrix of a graph (just "the Laplacian" for short) is defined as the *degree matrix*, *D*, which contains the degree of each node along the diagonal and zero everywhere else, minus the adjacency matrix *A*:

$$L = D - A$$

We definitely can't fit all of the linear algebra theory needed to understand the properties of this matrix, but suffice it to say: it has some *great* properties. We will exploit a couple in the following paragraphs.

First, we will look at the *eigenvectors* of L. An eigenvector v of a matrix M is a vector that satisfies the property $Mv = \lambda v$ for some number λ, known as the eigenvalue. In other words, v is a special vector in relation to M because Mv simply changes the size of the vector, without changing its direction. As we will soon see, eigenvectors have many useful properties—sometimes seeming even magical!

As an example, a 3×3 rotation matrix R, when multiplied to any 3D vector p, rotates it 30 degrees around the z-axis. R will rotate all vectors except for those that lie *on* the z-axis. For those, we'll see no effect, or $Rp = p$ (i.e., $Rp = \lambda p$) with eigenvalue $\lambda = 1$.

Exercise: Rotation Matrix

Consider this rotation matrix:

$$R = \begin{bmatrix} \cos\theta & -\sin\theta & 0 \\ \sin\theta & \cos\theta & 0 \\ 0 & 0 & 1 \end{bmatrix}$$

When R is multiplied with a 3D column-vector $p = [x\ y\ z]^T$, the resulting vector Rp is rotated by θ degrees around the z-axis.

1. For $\theta = 45°$, verify (by testing on a few arbitrary vectors) that R rotates these vectors around the z-axis. Remember that matrix multiplication in Python is denoted by @.

2. What does the matrix $S = RR$ do? Verify this in Python.

3. Verify that multiplying by R leaves the vector $[0\ 0\ 1]^T$ unchanged. In other words, $Rp = 1p$, which means p is an eigenvector of R with eigenvalue 1.

4. Use np.linalg.eig to find the eigenvalues and eigenvectors of R, and verify that $[0, 0, 1]^T$ is indeed among them, and that it corresponds to the eigenvalue 1.

Back to the Laplacian. A common problem in network analysis is visualization. How do you draw nodes and edges in such a way that you don't get a complete mess such as the one in Figure 6-1?

Figure 6-1. Visualization of Wikipedia structure (created by Chris Davis and released under CC-BY-SA-3.0 (http://bit.ly/2tj5tcA))

One way is to put nodes that share many edges close together. It turns out that we can do this by using the second-smallest eigenvalue of the Laplacian matrix, and its corresponding eigenvector, which is so important it has its own name: the Fiedler vector (*http://bit.ly/2tji13N*).

Let's use a minimal network to illustrate this. We start by creating the adjacency matrix:

```
import numpy as np
A = np.array([[0, 1, 1, 0, 0, 0],
              [1, 0, 1, 0, 0, 0],
              [1, 1, 0, 1, 0, 0],
              [0, 0, 1, 0, 1, 1],
```

```
                   [0, 0, 0, 1, 0, 1],
                   [0, 0, 0, 1, 1, 0]], dtype=float)
```

We can use NetworkX to draw this network. First, we initialize Matplotlib as usual:

```
# Make plots appear inline, set custom plotting style
%matplotlib inline
import matplotlib.pyplot as plt
plt.style.use('style/elegant.mplstyle')
```

Now, we can plot it:

```
import networkx as nx
g = nx.from_numpy_matrix(A)
layout = nx.spring_layout(g, pos=nx.circular_layout(g))
nx.draw(g, pos=layout,
        with_labels=True, node_color='white')
```

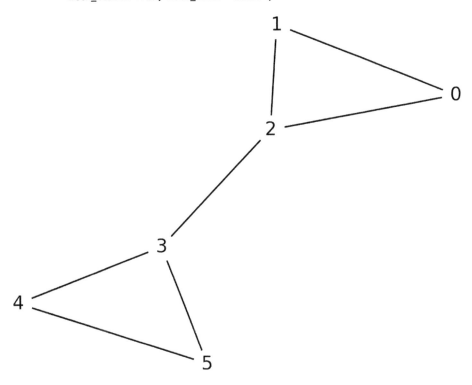

You can see that the nodes fall naturally into two groups, 0, 1, 2 and 3, 4, 5. Can the Fiedler vector tell us this? First, we must compute the degree matrix and the Laplacian. We first get the degrees by summing along either axis of A. (Either axis works because A is symmetric.)

```
d = np.sum(A, axis=0)
print(d)
```

```
[ 2.  2.  3.  3.  2.  2.]
```

We then put those degrees into a diagonal matrix of the same shape as A, the *degree matrix*. We can use the `np.diag` function to do this:

```
D = np.diag(d)
print(D)

[[ 2.  0.  0.  0.  0.  0.]
 [ 0.  2.  0.  0.  0.  0.]
 [ 0.  0.  3.  0.  0.  0.]
 [ 0.  0.  0.  3.  0.  0.]
 [ 0.  0.  0.  0.  2.  0.]
 [ 0.  0.  0.  0.  0.  2.]]
```

Finally, we get the Laplacian from the definition:

```
L = D - A
print(L)

[[ 2. -1. -1.  0.  0.  0.]
 [-1.  2. -1.  0.  0.  0.]
 [-1. -1.  3. -1.  0.  0.]
 [ 0.  0. -1.  3. -1. -1.]
 [ 0.  0.  0. -1.  2. -1.]
 [ 0.  0.  0. -1. -1.  2.]]
```

Because L is symmetric, we can use the `np.linalg.eigh` function to compute the eigenvalues and eigenvectors:

```
val, Vec = np.linalg.eigh(L)
```

You can verify that the values returned satisfy the definition of eigenvalues and eigenvectors. For example, one of the eigenvalues is 3:

```
np.any(np.isclose(val, 3))

True
```

And we can check that multiplying the matrix L by the corresponding eigenvector does indeed multiply the vector by 3:

```
idx_lambda3 = np.argmin(np.abs(val - 3))
v3 = Vec[:, idx_lambda3]

print(v3)
print(L @ v3)

[ 0.          0.37796447 -0.37796447 -0.37796447  0.68898224 -0.31101776]
[ 0.          1.13389342 -1.13389342 -1.13389342  2.06694671 -0.93305329]
```

As previously mentioned, the Fiedler vector is the vector corresponding to the second-smallest eigenvalue of L. Sorting the eigenvalues tells us which one is the second-smallest:

```
plt.plot(np.sort(val), linestyle='-', marker='o');
```

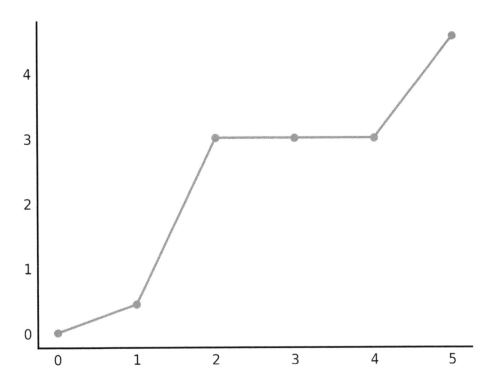

This is the first nonzero eigenvalue, close to 0.4. The Fiedler vector is the corresponding eigenvector (see Figure 6-2):

```
f = Vec[:, np.argsort(val)[1]]
plt.plot(f, linestyle='-', marker='o');
```

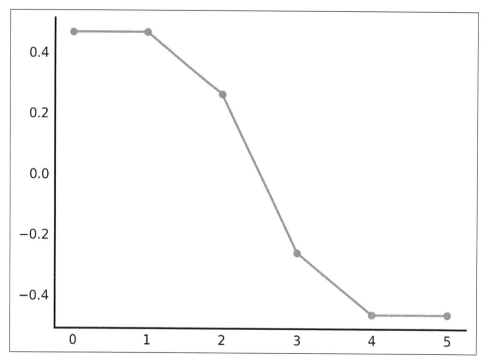

Figure 6-2. Fiedler vector of L

It's pretty remarkable: by looking at the *sign* of the elements of the Fiedler vector, we can separate the nodes into the two groups we identified in the drawing (see Figure 6-3)!

```
colors = ['orange' if eigv > 0 else 'gray' for eigv in f]
nx.draw(g, pos=layout, with_labels=True, node_color=colors)
```

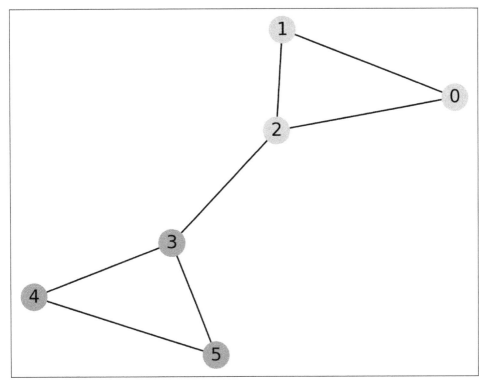

Figure 6-3. Nodes colored by their sign in the Fiedler vector of L

Laplacians with Brain Data

Let's demonstrate this process in a real-world example by laying out the brain cells in a worm, as shown in Figure 2 (*http://bit.ly/2s9unuL*) from the Varshney et al. paper (*http://bit.ly/2s9unuL*) that we introduced in Chapter 3. (Information on how to do this is in the supplementary material (*http://bit.ly/2sdZLIK*) for the paper.) To obtain their layout of the worm brain neurons, they used a related matrix, called the *degree-normalized Laplacian*.

Because the order of the neurons is important in this analysis, we will use a prepro-cessed dataset, rather than clutter this chapter with data cleaning. We got the original data from Lav Varshney's website (*http://www.ifp.illinois.edu/~varshney/elegans*), and the processed data is in our *data/* directory.

First, let's load the data. There are four components:

- The network of chemical synapses, through which a *pre-synaptic neuron* sends a chemical signal to a *post-synaptic* neuron

- The gap junction network, which contains direct electrical contacts between neurons

- The neuron IDs (names)

- The three neuron types:

  — *Sensory neurons*, those that detect signals coming from the outside world, encoded as 0

  — *Motor neurons*, those that activate muscles, enabling the worm to move, encoded as 2

  — *Interneurons*, the neurons in between, which enable complex signal processing between sensory neurons and motor neurons, encoded as 1

```python
import numpy as np
Chem = np.load('data/chem-network.npy')
Gap = np.load('data/gap-network.npy')
neuron_ids = np.load('data/neurons.npy')
neuron_types = np.load('data/neuron-types.npy')
```

We then simplify the network, adding the two kinds of connections together, and removing the directionality of the network by taking the average of in-connections and out-connections of neurons. This seems a bit like cheating but, since we are only looking for the *layout* of the neurons on a graph, we only care about *whether* neurons are connected, not in which direction. We are going to call the resulting matrix the *connectivity* matrix, C, which is just a different kind of adjacency matrix.

```python
A = Chem + Gap
C = (A + A.T) / 2
```

To get the Laplacian matrix L, we need the degree matrix D, which contains the degree of node i at position [i, i], and zeros everywhere else.

```python
degrees = np.sum(C, axis=0)
D = np.diag(degrees)
```

Now, we can get the Laplacian just like before:

```python
L = D - C
```

The vertical coordinates in Figure 2 (*http://bit.ly/2s9unuL*) from the paper are given by arranging nodes such that, on average, neurons are as close as possible to "just above" their downstream neighbors. Varshney et al. call this measure "processing depth," and it's obtained by solving a linear equation involving the Laplacian. We use scipy.linalg.pinv, the pseudoinverse (*http://bit.ly/2tqOJQY*), to solve it:

```python
from scipy import linalg
b = np.sum(C * np.sign(A - A.T), axis=1)
z = linalg.pinv(L) @ b
```

(Note the use of the @ symbol, which was introduced in Python 3.5 to denote matrix multiplication. As we noted in the preface and in Chapter 5, in previous versions of Python, you need to use the function `np.dot`.)

In order to obtain the degree-normalized Laplacian, Q, we need the inverse square root of the D matrix:

```
Dinv2 = np.diag(1 / np.sqrt(degrees))
Q = Dinv2 @ L @ Dinv2
```

Finally, we are able to extract the x coordinates of the neurons to ensure that highly connected neurons remain close: the eigenvector of Q corresponding to its second-smallest eigenvalue, normalized by the degrees:

```
val, Vec = linalg.eig(Q)
```

Note from the documentation of `numpy.linalg.eig`:

> The eigenvalues are not necessarily ordered.

Although the documentation in SciPy's `eig` lacks this warning, it remains true in this case. We must therefore sort the eigenvalues and the corresponding eigenvector columns ourselves:

```
smallest_first = np.argsort(val)
val = val[smallest_first]
Vec = Vec[:, smallest_first]
```

Now we can find the eigenvector we need to compute the affinity coordinates:

```
x = Dinv2 @ Vec[:, 1]
```

(The reasons for using this vector are too long to explain here, but appear in the paper's supplementary material, linked above. The short version is that choosing this vector minimizes the total length of the links between neurons.)

There is one small kink that we must address before proceeding: eigenvectors are only defined up to a multiplicative constant. This follows simply from the definition of an eigenvector: suppose v is an eigenvector of the matrix M, with corresponding eigenvalue λ. Then αv is also an eigenvector of M for any scalar number α, because $Mv = \lambda v$ implies $M(\alpha v) = \lambda(\alpha v)$. So, it is arbitrary whether a software package returns v or $-v$ when asked for the eigenvectors of M. In order to make sure we reproduce the layout from the Varshney et al. paper, we must make sure that the vector is pointing in the same direction as theirs, rather than the opposite direction. We do this by choosing an arbitrary neuron from their Figure 2, and checking the sign of x at that position. We then reverse it if it doesn't match its sign in Figure 2 of the paper.

```
vc2_index = np.argwhere(neuron_ids == 'VC02')
if x[vc2_index] < 0:
    x = -x
```

Now it's just a matter of drawing the nodes and the edges. We color them according to the type stored in neuron_types using the appealing and functional "colorblind" colorbrewer palette (*http://chrisalbon.com/python/seaborn_color_palettes.html*):

```python
from matplotlib.colors import ListedColormap
from matplotlib.collections import LineCollection


def plot_connectome(x_coords, y_coords, conn_matrix, *,
                    labels=(), types=None, type_names=('',),
                    xlabel='', ylabel=''):
    """Plot neurons as points connected by lines.

    Neurons can have different types (up to 6 distinct colors).

    Parameters
    ----------
    x_coords, y_coords : array of float, shape (N,)
        The x-coordinates and y-coordinates of the neurons.
    conn_matrix : array or sparse matrix of float, shape (N, N)
        The connectivity matrix, with nonzero entry (i, j) if and only
        if node i and node j are connected.
    labels : array-like of string, shape (N,), optional
        The names of the nodes.
    types : array of int, shape (N,), optional
        The type (e.g. sensory neuron, interneuron) of each node.
    type_names : array-like of string, optional
        The name of each value of `types`. For example, if a 0 in
        `types` means "sensory neuron", then `type_names[0]` should
        be "sensory neuron".
    xlabel, ylabel : str, optional
        Labels for the axes.
    """
    if types is None:
        types = np.zeros(x_coords.shape, dtype=int)
    ntypes = len(np.unique(types))
    colors = plt.rcParams['axes.prop_cycle'][:ntypes].by_key()['color']
    cmap = ListedColormap(colors)

    fig, ax = plt.subplots()

    # plot neuron locations:
    for neuron_type in range(ntypes):
        plotting = (types == neuron_type)
        pts = ax.scatter(x_coords[plotting], y_coords[plotting],
                         c=cmap(neuron_type), s=4, zorder=1)
        pts.set_label(type_names[neuron_type])

    # add text labels:
    for x, y, label in zip(x_coords, y_coords, labels):
        ax.text(x, y, '   ' + label,
                verticalalignment='center', fontsize=3, zorder=2)
```

```
# plot edges
pre, post = np.nonzero(conn_matrix)
links = np.array([[x_coords[pre], x_coords[post]],
                  [y_coords[pre], y_coords[post]]]).T
ax.add_collection(LineCollection(links, color='lightgray',
                                 lw=0.3, alpha=0.5, zorder=0))

ax.legend(scatterpoints=3, fontsize=6)

ax.set_xlabel(xlabel, fontsize=8)
ax.set_ylabel(ylabel, fontsize=8)

plt.show()
```

Now, let's use that function to plot the neurons:

```
plot_connectome(x, z, C, labels=neuron_ids, types=neuron_types,
                type_names=['sensory neurons', 'interneurons',
                            'motor neurons'],
                xlabel='Affinity eigenvector 1', ylabel='Processing depth')
```

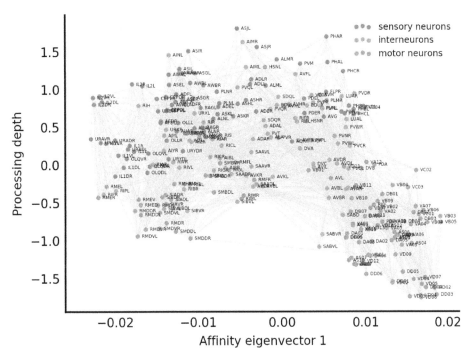

There you are: a worm brain! As discussed in the original paper, you can see the top-down processing from sensory neurons to motor neurons through a network of interneurons. You can also see two distinct groups of motor neurons: these correspond to the neck (left) and body (right) body segments of the worm.

Exercise: Showing the Affinity View

How do you modify the above code to show the affinity view in Figure 2B from the paper?

Exercise Challenge: Linear Algebra with Sparse Matrices

The preceding code uses NumPy arrays to hold the matrix and perform the necessary computations. Because we are using a small graph of fewer than 300 nodes, this is feasible. However, for larger graphs, it would fail.

For example, one might want to analyze the relationships between libraries listed on the Python Package Index, or PyPI, which contains over 100,00 packages. Holding the Laplacian matrix for this graph would take up $8(100 \times 10^3)^2 = 8 \times 10^{10}$ bytes, or 80 GB, of RAM. If you add to that the adjacency, symmetric adjacency, pseudoinverse, and, say, two temporary matrices used during calculations, you climb up to 480 GB, beyond the reach of most desktop computers.

"Ha!" some of you might be thinking. "Ha! My desktop has 512 GB of RAM! It would make short work of this so-called 'large' graph!"

Perhaps. But you might also want to analyze the Association for Computing Machinery (ACM) citation graph, a network of over two million scholarly works and references. *That* Laplacian would take up 32 terabytes of RAM.

However, we know that the dependency and reference graphs are *sparse*: packages usually depend on just a few other packages, not on the whole of PyPI. And papers and books usually only reference a few others, too. So we can hold the above matrices using the sparse data structures from `scipy.sparse` (see Chapter 5), and use the linear algebra functions in `scipy.sparse.linalg` to compute the values we need.

Try to explore the documentation in `scipy.sparse.linalg` to come up with a sparse version of the above computation.

> The pseudoinverse of a sparse matrix is, in general, not sparse, so you can't use it here. Similarly, you can't get all the eigenvectors of a sparse matrix, because they would together make up a dense matrix.

You'll find parts of the solution below (and of course in the Appendix), but we highly recommend that you try it out on your own.

Solvers

SciPy has several sparse iterative solvers available, and it is not always obvious which to use. Unfortunately, that question also has no easy answer: different algorithms have different strengths in terms of speed of convergence, stability, accuracy, and memory use (among others). It is also not possible to predict, by looking at the input data, which algorithm will perform best.

Here is a rough guideline for choosing an iterative solver:

- If A, the input matrix, is symmetric and positive definite, use the conjugate gradient solver `cg`. If A is symmetric, but near-singular or indefinite, try the minimum residual iteration method `minres`.
- For nonsymmetric systems, try the biconjugate gradient stabilized method, `bicg stab`. The conjugate gradient squared method, `cgs`, is a bit faster, but has more erratic convergence.
- If you need to solve many similar systems, use the LGMRES algorithm `lgmres`.
- If A is not square, use the least squares algorithm `lsmr`.

For further reading, see:

- Noël M. Nachtigal, Satish C. Reddy, and Lloyd N. Trefethen, "How Fast Are Non-symmetric Matrix Iterations?" *SIAM Journal on Matrix Analysis and Applications*13, no. 3 (1992): 778–95. 778-795.
- Jack Dongarra, "Survey of Recent Krylov Methods" (*http://www.netlib.org/linalg/html_templates/node50.html*)," November 20, 1995.

PageRank: Linear Algebra for Reputation and Importance

Another application of linear algebra and eigenvectors is Google's PageRank algorithm, which is punnily named both for web pages and for one of its cofounders, Larry Page.

To rank web pages by importance, you might count how many other web pages link to it. After all, if everyone is linking to a particular page, it must be good, right? But this metric is easily gamed: to make your own web page rise in the rankings, just create as many other web pages as you can and have them all link to your original page.

The key insight that drove Google's early success was that important web pages are not linked to by just many web pages, but by *important* web pages. And how do we know that those other pages are important? Because they themselves are linked to by important pages. And so on.

This recursive definition implies that page importance can be measured by an eigenvector of the so-called *transition matrix*, which contains the links between web pages. Suppose you have your vector of importance r, and your matrix of links M. You don't know r yet, but you do know that the importance of a page is proportional to the sum of importances of the pages that link to it: $r = \alpha Mr$, or $Mr = \lambda r$, for $\lambda = 1/\alpha$. That's just the definition of an eigenvalue!

By ensuring some special properties are satisfied by the transition matrix, we can further determine that the required eigenvalue is 1, and that it is the largest eigenvalue of M.

The transition matrix imagines a web surfer, often named Webster, randomly clicking a link from each web page he visits, and then asks, what's the probability that he ends up at any given page? This probability is called the PageRank.

Since Google's rise, researchers have been applying PageRank to all sorts of networks. We'll use an example by Stefano Allesina and Mercedes Pascual, which they published (*https://doi.org/10.1371/journal.pcbi.1000494*) in *PLoS Computational Biology*. They thought to apply the method in ecological *food webs*, networks that link species to those that they eat.

Naively, if you wanted to see how critical a species was for an ecosystem, you would look at how many species eat it. If it's many, and that species disappeared, then all its "dependent" species might disappear with it. In network parlance, you could say that its *in-degree* determines its ecological importance.

Could PageRank be a better measure of importance for an ecosystem?

Professor Allesina kindly provided us with a few food webs to play around with. We've saved one of these, from the St. Marks National Wildlife Refuge in Florida, in the Graph Markup Language format. The web was described (*http://bit.ly/2sdWJEc*) in 1999 by Robert R. Christian and Joseph J. Luczovich. In the dataset, a node i has an edge to node j if species i eats species j.

We'll start by loading in the data, which NetworkX knows how to read trivially:

```
import networkx as nx

stmarks = nx.read_gml('data/stmarks.gml')
```

Next, we get the sparse matrix corresponding to the graph. Because a matrix only holds numerical information, we need to maintain a separate list of package names corresponding to the matrix rows/columns:

```
species = np.array(stmarks.nodes())  # array for multiindexing
Adj = nx.to_scipy_sparse_matrix(stmarks, dtype=np.float64)
```

From the adjacency matrix, we can derive a *transition probability* matrix, where every edge is replaced by a *probability* of 1 over the number of outgoing edges from that species. In the food web, it might make more sense to call this a lunch probability matrix.

The total number of species in our matrix is going to be used a lot, so let's call it n:

```
n = len(species)
```

Next, we need the degrees, and, in particular, the *diagonal matrix* containing the inverse of the out-degrees of each node on the diagonal:

```
np.seterr(divide='ignore')  # ignore division-by-zero errors
from scipy import sparse

degrees = np.ravel(Adj.sum(axis=1))
Deginv = sparse.diags(1 / degrees).tocsr()

Trans = (Deginv @ Adj).T
```

Normally, the PageRank score would simply be the first eigenvector of the transition matrix. If we call the transition matrix M and the vector of PageRank values r, we have:

$$r = Mr$$

But the `np.seterr` call is a clue that it's not quite so simple. The PageRank approach only works when the transition matrix is a *column-stochastic* matrix, in which every column sums to 1. Additionally, every page must be reachable from every other page, even if the path to reach it is very long.

In our food web, this causes problems, because the bottom of the food chain, what the authors call *detritus* (basically sea sludge), doesn't actually *eat* anything (the Circle of Life notwithstanding), so you can't reach other species from it.

> *Young Simba:* But, Dad, don't we eat the antelope?
>
> *Mufasa:* Yes, Simba, but let me explain. When we die, our bodies become the grass, and the antelope eat the grass. And so we are all connected in the great Circle of Life.
>
> —*The Lion King*

To deal with this, the PageRank algorithm uses a so-called "damping factor," usually taken to be 0.85. This means that 85% of the time, the algorithm follows a link at random, but for the other 15%, it randomly jumps to any arbitrary page. It's as if every page had a low probability link to every other page. Or, in our case, it's as if shrimp, on rare occasions, ate sharks. It might seem nonsensical, but bear with us! It is, in fact, the mathematical representation of the Circle of Life. We'll set it to 0.85, but actually it doesn't really matter for this analysis: the results are similar for a large range of possible damping factors.

If we call the damping factor d, then the modified PageRank equation is:

$$r = dMr + \frac{1-d}{n}\mathbf{1}$$

and:

$$(\boldsymbol{I} - dM)r = \frac{1-d}{n}\mathbf{1}$$

We can solve this equation using `scipy.sparse.linalg`'s direct solver, `spsolve`. Depending on the structure and size of a linear algebra problem, though, it might be more efficient to use an iterative solver. See the `scipy.sparse.linalg` documentation (*http://bit.ly/2se21Qg*) for more information on this.

```python
from scipy.sparse.linalg import spsolve

damping = 0.85
beta = 1 - damping

I = sparse.eye(n, format='csc')  # Same sparse format as Trans

pagerank = spsolve(I - damping * Trans,
                   np.full(n, beta / n))
```

We now have the "foodrank" of the St. Marks food web!

So how does a species' foodrank compare to the number of other species eating it?

```python
def pagerank_plot(in_degrees, pageranks, names, *,
                  annotations=[], **figkwargs):
    """Plot node pagerank against in-degree, with hand-picked node names."""

    fig, ax = plt.subplots(**figkwargs)
    ax.scatter(in_degrees, pageranks, c=[0.835, 0.369, 0], lw=0)
    for name, indeg, pr in zip(names, in_degrees, pageranks):
        if name in annotations:
            text = ax.text(indeg + 0.1, pr, name)

    ax.set_ylim(0, np.max(pageranks) * 1.1)
    ax.set_xlim(-1, np.max(in_degrees) * 1.1)
    ax.set_ylabel('PageRank')
    ax.set_xlabel('In-degree')
```

We now draw the plot. Having explored the dataset before writing this, we have pre-labeled some interesting nodes in the plot:

```python
interesting = ['detritus', 'phytoplankton', 'benthic algae', 'micro-epiphytes',
               'microfauna', 'zooplankton', 'predatory shrimps', 'meiofauna',
               'gulls']
in_degrees = np.ravel(Adj.sum(axis=0))
pagerank_plot(in_degrees, pagerank, species, annotations=interesting)
```

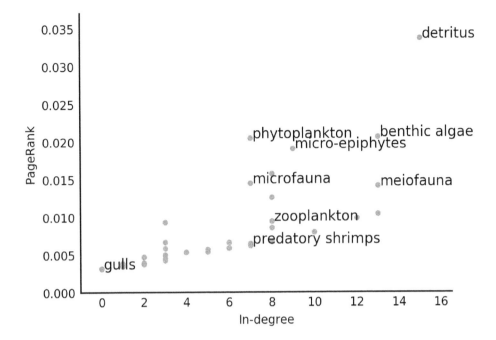

Sea sludge (detritus) is the most important element both by number of species feeding on it (15) and by PageRank (>0.003). But the second most important element is *not* benthic algae, which feeds 13 other species, but rather phytoplankton, which feeds just 7! That's because other *important* species feed on it. On the bottom left, we've got sea gulls, who, we can now confirm, do bugger-all for the ecosystem. Those vicious *predatory shrimps* (we're not making this up) support the same number of species as phytoplankton, but they are less essential species, so they end up with a lower foodrank.

Although we won't do it here, Allesina and Pascual go on to model the ecological impact of species extinction, and indeed find that PageRank predicts ecological importance better than in-degree.

Before we wrap up though, we'll note that PageRank can be computed several different ways. One way, complementary to what we did above, is called the *power method*, and it's quite, well, powerful! It stems from the Perron-Frobenius theorem (*http:// bit.ly/2seyshv*), which states, among other things, that a stochastic matrix has 1 as an eigenvalue, and that this is its *largest* eigenvalue. (The corresponding eigenvector is the PageRank vector.) What this means is that, whenever we multiply *any* vector by *M*, its component pointing toward this major eigenvector stays the same, while *all other components shrink* by a multiplicative factor. The consequence is that if we multiply some random starting vector by *M* repeatedly, we should eventually get the PageRank vector!

SciPy makes this very efficient with its sparse matrix module:

```
def power(Trans, damping=0.85, max_iter=10**5):
    n = Trans.shape[0]
    r0 = np.full(n, 1/n)
    r = r0
    for _iter_num in range(max_iter):
        rnext = damping * Trans @ r + (1 - damping) / n
        if np.allclose(rnext, r):
            break
        r = rnext
    return r
```

Exercise: Dealing with Dangling Nodes

In the preceding iteration, note that Trans is *not* column-stochastic, so the r vector gets shrunk at each iteration. In order to make the matrix stochastic, we have to replace every zero-column by a column of all $1/n$. This is too expensive, but computing the iteration is cheaper. How can you modify the code above to ensure that r remains a probability vector throughout?

Exercise: Equivalence of Different Eigenvector Methods

Verify that these three methods all give the same ranking for the nodes. numpy.corr coef might be a useful function for this.

Concluding Remarks

The field of linear algebra is far too broad to adequately cover in a chapter, but this chapter gave you a glimpse into its power, and of the way Python, NumPy, and SciPy make its elegant algorithms accessible.

## CHAPTER 7

Function Optimization in SciPy

"What's new?" is an interesting and broadening eternal question, but one which, if pursued exclusively, results only in an endless parade of trivia and fashion, the silt of tomorrow. I would like, instead, to be concerned with the question "What is best?", a question which cuts deeply rather than broadly, a question whose answers tend to move the silt downstream.

—Robert M. Pirsig, *Zen and the Art of Motorcycle Maintenance*

When hanging a picture on the wall, it is sometimes difficult to get it straight. You make an adjustment, step back, evaluate the picture's horizontality, and repeat. This is a process of *optimization*: we're changing the orientation of the portrait until it satisfies our demand—that it makes a zero angle with the horizon.

In mathematics, our demand is called a "cost function," and the orientation of the portrait the "parameter." In a typical optimization problem, we vary the parameters until the cost function is minimized.

Consider, for example, the shifted parabola, $f(x) = (x - 3)^2$. We'd like to find the value of x that minimizes this cost function. We know that this function, with parameter x, has a minimum at 3, because we can calculate the derivative, set it to zero, and see that $2(x - 3) = 0$ (i.e., $x = 3$).

But, if this function were much more complicated (e.g., if the expression had many terms, multiple points of zero derivative, contained nonlinearities, or was dependent on more variables), using manual calculation would become arduous.

You can think of the cost function as representing a landscape, where we are trying to find the lowest point. That analogy immediately highlights one of the hard parts of this problem: if you are standing in any valley, with mountains surrounding you, how do you know whether you are in the lowest valley, or whether this valley just seems low because it is surrounded by particularly tall mountains? In optimization parlance:

how do you know whether you are trapped in a *local minimum*? Most optimization algorithms make some attempt to address the issue.[1]

 Figure 7-1 shows all of the methods that are available in SciPy, some of which we will use, while others we'll leave you to discover.

There are many different optimization algorithms to choose from (see Figure 7-1). You get to choose whether your cost function takes a scalar or a vector as input (i.e., do you have one or multiple parameters to optimize?). There are those that require the cost function gradient to be given and those that automatically estimate it. Some only search for parameters in a given area (*constrained optimization*), and others examine the entire parameter space.

---

1 Optimization algorithms handle this issue in various ways, but two common approaches are line searches and trust regions. With a *line search*, you try to find the cost function minimum along a specific dimension, and then successively attempt the same along the other dimensions. With *trust regions*, we move our guess for the minimum in the direction we expect it to be; if we see that we are indeed approaching the minimum as expected, we repeat the procedure with increased confidence. If not, we lower our confidence and search a wider area.

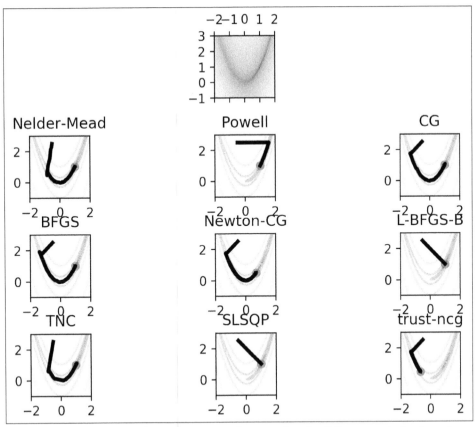

Figure 7-1. Comparison of optimization pathways taken by different optimization algorithms on the Rosenbrock function (top). Powell's method performs a line search along the first dimension before doing gradient descent. The conjugate gradient (CG) method, on the other hand, performs gradient descent from the starting point

Optimization in SciPy: scipy.optimize

In the rest of this chapter, we are going to use SciPy's `optimize` module to align two images. Applications of image alignment, or *registration*, include panorama stitching, combining different brain scans, super-resolution imaging, and, in astronomy, object denoising (noise reduction) through the combination of multiple exposures.

As usual, we set up our plotting environment:

```
# Make plots appear inline, set custom plotting style
%matplotlib inline
import matplotlib.pyplot as plt
plt.style.use('style/elegant.mplstyle')
```

Let's start with the simplest version of the problem: we have two images, one shifted relative to the other. We wish to recover the shift that will best align our images.

Our optimization function will "jiggle" one of the images, and see whether jiggling it in one direction or another reduces their dissimilarity. By doing this repeatedly, we can try to find the correct alignment.

An Example: Computing Optimal Image Shift

You'll remember our astronaut—Eileen Collins—from Chapter 3. We will be shifting this image by 50 pixels to the right, then comparing it to the original until we find the shift that best matches. Obviously this is a silly thing to do, as we know the original position, but this way we know the truth, and we can check how our algorithm is doing. Here's the original and shifted image:

```
from skimage import data, color
from scipy import ndimage as ndi

astronaut = color.rgb2gray(data.astronaut())
shifted = ndi.shift(astronaut, (0, 50))

fig, axes = plt.subplots(nrows=1, ncols=2)
axes[0].imshow(astronaut)
axes[0].set_title('Original')
axes[1].imshow(shifted)
axes[1].set_title('Shifted');
```

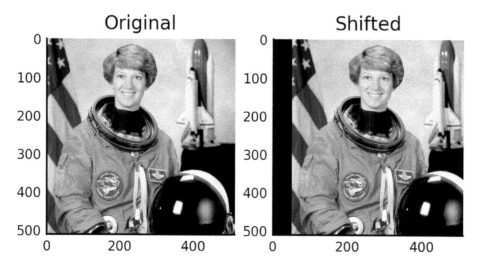

For the optimization algorithm to do its work, we need some way of defining "dissimilarity"—that is, the cost function. The easiest way to do this is to simply calculate the average of the squared differences, often called the *mean squared error*, or MSE.

```
import numpy as np

def mse(arr1, arr2):
    """Compute the mean squared error between two arrays."""
    return np.mean((arr1 - arr2)**2)
```

This will return 0 when the images are perfectly aligned, and a higher value otherwise. With this cost function, we can check whether two images are aligned:

```
ncol = astronaut.shape[1]

# Cover a distance of 90% of the length in columns,
# with one value per percentage point
shifts = np.linspace(-0.9 * ncol, 0.9 * ncol, 181)
mse_costs = []

for shift in shifts:
    shifted_back = ndi.shift(shifted, (0, shift))
    mse_costs.append(mse(astronaut, shifted_back))

fig, ax = plt.subplots()
ax.plot(shifts, mse_costs)
ax.set_xlabel('Shift')
ax.set_ylabel('MSE');
```

With the cost function defined, we can ask `scipy.optimize.minimize` to search for optimal parameters:

```
from scipy import optimize

def astronaut_shift_error(shift, image):
    corrected = ndi.shift(image, (0, shift))
    return mse(astronaut, corrected)

res = optimize.minimize(astronaut_shift_error, 0, args=(shifted,),
                        method='Powell')

print(f'The optimal shift for correction is: {res.x}')

The optimal shift for correction is: -49.99997565757551
```

It worked! We shifted it by +50 pixels, and, thanks to our MSE measure, SciPy's `optimize.minimize` function has given us the correct amount of shift (–50) to get it back to its original state.

It turns out, however, that this was a particularly easy optimization problem, which brings us to the principal difficulty of this kind of alignment: sometimes, the MSE has to get worse before it gets better.

Let's look again at shifting images, starting with the unmodified image:

```
ncol = astronaut.shape[1]

# Cover a distance of 90% of the length in columns,
# with one value per percentage point
shifts = np.linspace(-0.9 * ncol, 0.9 * ncol, 181)
mse_costs = []

for shift in shifts:
    shifted1 = ndi.shift(astronaut, (0, shift))
    mse_costs.append(mse(astronaut, shifted1))

fig, ax = plt.subplots()
ax.plot(shifts, mse_costs)
ax.set_xlabel('Shift')
ax.set_ylabel('MSE');
```

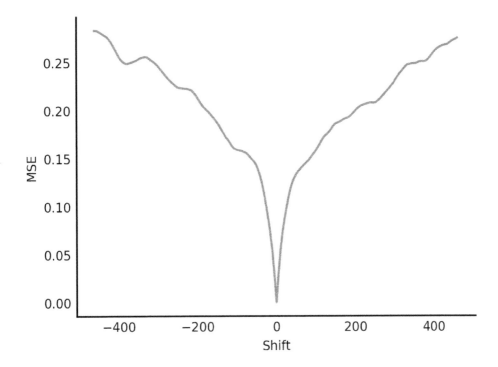

Starting at zero shift, have a look at the MSE value as the shift becomes increasingly negative: it increases consistently until around –300 pixels of shift, where it starts to decrease again! Only slightly, but it decreases nonetheless. The MSE bottoms out at around –400 before it increases again. This is called a *local minimum*. Because optimization methods only have access to "nearby" values of the cost function, if the function improves by moving in the "wrong" direction, the minimize process will move that way regardless. So, if we start by an image shifted by –340 pixels:

```
shifted2 = ndi.shift(astronaut, (0, -340))
```

minimize will shift it by a further 40 pixels or so, instead of recovering the original image:

```
res = optimize.minimize(astronaut_shift_error, 0, args=(shifted2,),
                        method='Powell')

print(f'The optimal shift for correction is {res.x}')

The optimal shift for correction is -38.51778619397471
```

The common solution to this problem is to smooth or downscale the images, which has the dual result of smoothing the objective function. Have a look at the same plot, after having smoothed the images with a Gaussian filter:

```
from skimage import filters

astronaut_smooth = filters.gaussian(astronaut, sigma=20)

mse_costs_smooth = []
shifts = np.linspace(-0.9 * ncol, 0.9 * ncol, 181)
for shift in shifts:
    shifted3 = ndi.shift(astronaut_smooth, (0, shift))
    mse_costs_smooth.append(mse(astronaut_smooth, shifted3))

fig, ax = plt.subplots()
ax.plot(shifts, mse_costs, label='original')
ax.plot(shifts, mse_costs_smooth, label='smoothed')
ax.legend(loc='lower right')
ax.set_xlabel('Shift')
ax.set_ylabel('MSE');
```

As you can see, with some rather extreme smoothing, the "funnel" of the error function becomes wider and less bumpy. Rather than smoothing the function itself, we can get a similar effect by blurring the images before comparing them. Therefore, modern alignment software uses what's called a *Gaussian pyramid*, which is a set of progressively lower-resolution versions of the same image. We align the lower-resolution (blurrier) images first, to get an approximate alignment, and then progressively refine the alignment with sharper images.

```
def downsample2x(image):
    offsets = [((s + 1) % 2) / 2 for s in image.shape]
    slices = [slice(offset, end, 2)
                for offset, end in zip(offsets, image.shape)]
    coords = np.mgrid[slices]
    return ndi.map_coordinates(image, coords, order=1)


def gaussian_pyramid(image, levels=6):
    """Make a Gaussian image pyramid.

    Parameters
    ----------
    image : array of float
        The input image.
    max_layer : int, optional
        The number of levels in the pyramid.

    Returns
    -------
    pyramid : iterator of array of float
        An iterator of Gaussian pyramid levels, starting with the top
        (lowest resolution) level.
    """
    pyramid = [image]

    for level in range(levels - 1):
        blurred = ndi.gaussian_filter(image, sigma=2/3)
        image = downsample2x(image)
        pyramid.append(image)

    return reversed(pyramid)
```

Let's see how the 1D alignment looks along that pyramid:

```
shifts = np.linspace(-0.9 * ncol, 0.9 * ncol, 181)
nlevels = 8
costs = np.empty((nlevels, len(shifts)), dtype=float)
astronaut_pyramid = list(gaussian_pyramid(astronaut, levels=nlevels))
for col, shift in enumerate(shifts):
    shifted = ndi.shift(astronaut, (0, shift))
    shifted_pyramid = gaussian_pyramid(shifted, levels=nlevels)
    for row, image in enumerate(shifted_pyramid):
        costs[row, col] = mse(astronaut_pyramid[row], image)

fig, ax = plt.subplots()
for level, cost in enumerate(costs):
    ax.plot(shifts, cost, label='Level %d' % (nlevels - level))
ax.legend(loc='lower right', frameon=True, framealpha=0.9)
ax.set_xlabel('Shift')
ax.set_ylabel('MSE');
```

As you can see, at the highest level of the pyramid, that bump at a shift of about –325 disappears. We can therefore get an approximate alignment at that level, then pop down to the lower levels to refine that alignment (see Figure 7-2).

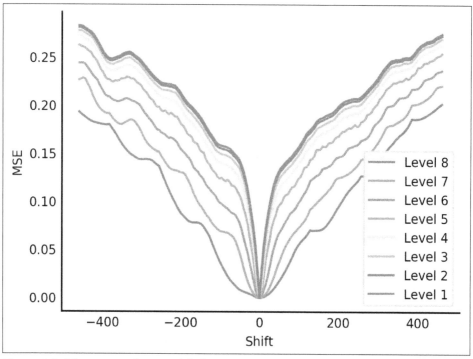

Figure 7-2. Mean squared error of shift at various levels of a Gaussian pyramid

Image Registration with Optimize

Let's automate that, and try with a "real" alignment, with three parameters: rotation, translation in the row dimension, and translation in the column dimension. This is called a *"rigid* registration" because there are no deformations of any kind (scaling, skew, or other stretching). The object is considered solid and is moved around (including rotation) until a match is found.

To simplify the code, we'll use the scikit-image `transform` module to compute the shift and rotation of the image. SciPy's `optimize` requires a vector of parameters as input. We first make a function that will take such a vector and produce a rigid trans-formation with the right parameters:

```
from skimage import transform

def make_rigid_transform(param):
    r, tc, tr = param
    return transform.SimilarityTransform(rotation=r,
```

```
                              translation=(tc, tr))

rotated = transform.rotate(astronaut, 45)

fig, axes = plt.subplots(nrows=1, ncols=2)
axes[0].imshow(astronaut)
axes[0].set_title('Original')
axes[1].imshow(rotated)
axes[1].set_title('Rotated');
```

Next, we need a cost function. This is just MSE, but SciPy requires a specific format:
the first argument needs to be the *parameter vector*, which it is optimizing. Subsequent arguments can be passed through the args keyword as a tuple, but must
remain fixed: only the parameter vector can be optimized. In our case, this is just the
rotation angle and the two translation parameters:

```
def cost_mse(param, reference_image, target_image):
    transformation = make_rigid_transform(param)
    transformed = transform.warp(target_image, transformation, order=3)
    return mse(reference_image, transformed)
```

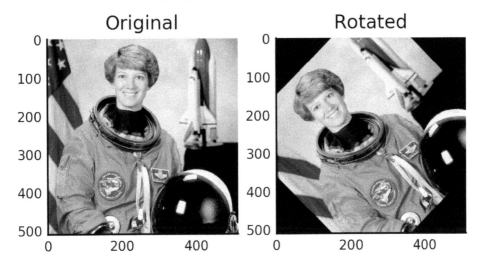

Finally, we write our alignment function, which optimizes our cost function *at each
level of the Gaussian pyramid*, using the result of the previous level as a starting point
for the next one:

```
def align(reference, target, cost=cost_mse):
    nlevels = 7
    pyramid_ref = gaussian_pyramid(reference, levels=nlevels)
    pyramid_tgt = gaussian_pyramid(target, levels=nlevels)

    levels = range(nlevels, 0, -1)
    image_pairs = zip(pyramid_ref, pyramid_tgt)
```

```
p = np.zeros(3)

for n, (ref, tgt) in zip(levels, image_pairs):
    p[1:] *= 2

    res = optimize.minimize(cost, p, args=(ref, tgt), method='Powell')
    p = res.x

    # print current level, overwriting each time (like a progress bar)
    print(f'Level: {n}, Angle: {np.rad2deg(res.x[0]) :.3}, '
          f'Offset: ({res.x[1] * 2**n :.3}, {res.x[2] * 2**n :.3}), '
          f'Cost: {res.fun :.3}', end='\r')

print('')  # newline when alignment complete
return make_rigid_transform(p)
```

Let's try it with our astronaut image. We rotate it by 60 degrees and add some noise to it. Can SciPy recover the correct transform? (See Figure 7-3.)

```
from skimage import util

theta = 60
rotated = transform.rotate(astronaut, theta)
rotated = util.random_noise(rotated, mode='gaussian',
                            seed=0, mean=0, var=1e-3)

tf = align(astronaut, rotated)
corrected = transform.warp(rotated, tf, order=3)

f, (ax0, ax1, ax2) = plt.subplots(1, 3)
ax0.imshow(astronaut)
ax0.set_title('Original')
ax1.imshow(rotated)
ax1.set_title('Rotated')
ax2.imshow(corrected)
ax2.set_title('Registered')
for ax in (ax0, ax1, ax2):
    ax.axis('off')

Level: 1, Angle: -60.0, Offset: (-1.87e+02, 6.98e+02), Cost: 0.0369
```

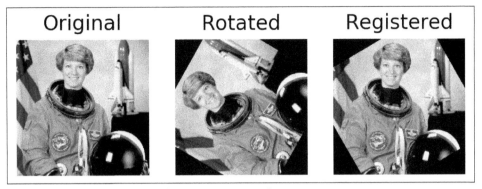

Figure 7-3. Optimization used to recover image alignment

We're feeling pretty good now. But our choice of parameters actually masked the difficulty of optimization. Let's see what happens with a rotation of 50 degrees, which is *closer* to the original image:

```
theta = 50
rotated = transform.rotate(astronaut, theta)
rotated = util.random_noise(rotated, mode='gaussian',
                            seed=0, mean=0, var=1e-3)

tf = align(astronaut, rotated)
corrected = transform.warp(rotated, tf, order=3)

f, (ax0, ax1, ax2) = plt.subplots(1, 3)
ax0.imshow(astronaut)
ax0.set_title('Original')
ax1.imshow(rotated)
ax1.set_title('Rotated')
ax2.imshow(corrected)
ax2.set_title('Registered')
for ax in (ax0, ax1, ax2):
    ax.axis('off')
```

```
Level: 1, Angle: 0.414, Offset: (2.85, 38.4), Cost: 0.141
```

Even though we started closer to the original image, we failed to recover the correct rotation (Figure 7-4). This is because optimization techniques can get stuck in local minima, little bumps on the road to success, as we saw above with the shift-only alignment. They can therefore be quite sensitive to the starting parameters.

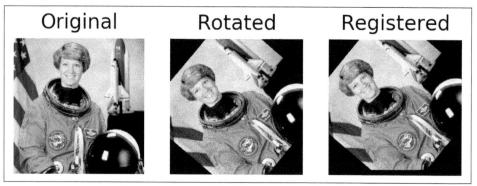

Figure 7-4. Failed optimization

Avoiding Local Minima with Basin Hopping

A 1997 algorithm devised by David Wales and Jonathan Doyle,[2] called *basin hopping*, attempts to avoid local minima by trying an optimization from some initial parameters, then moving away from the found local minimum in a random direction, and optimizing again. By choosing an appropriate step size for these random moves, the algorithm can avoid falling into the same local minimum twice, and thus explore a much larger area of the parameter space than simple gradient-based optimization methods.

We leave it as an exercise for readers to incorporate SciPy's implementation of basin hopping into our alignment function. You'll need it for later parts of the chapter, so feel free to peek at the solution at the end of the book if you're stuck.

Exercise: Modify the align Function

Try modifying the `align` function to use `scipy.optimize.basinhopping`, which has explicit strategies to avoid local minima.

> Limit using basin hopping to just the top levels of the pyramid, as it is a slower optimization approach, and could take rather long to run at full image resolution.

---

2 David J. Wales and Jonathan P.K. Doyle, "Global Optimization by Basin-Hopping and the Lowest Energy Structures of Lennard-Jones Clusters Containing up to 110 Atoms", *Journal of Physical Chemistry* 101, no. 28 (1997): 5111–16.

"What Is Best?": Choosing the Right Objective Function

At this point, we have a working registration approach, which is most excellent. But it turns out that we've only solved the easiest of registration problems: aligning images of the same *modality*. This means that we expect bright pixels in the reference image to match up to bright pixels in the test image.

We now move on to aligning different color channels of the same image, where we can no longer rely on the channels having the same modality. This task has historical significance: between 1909 and 1915, photographer Sergei Mikhailovich Prokudin-Gorskii produced color photographs of the Russian empire before color photography had been invented. He did this by taking three different monochrome pictures of a scene, each with a different color filter placed in front of the lens.

Aligning bright pixels together, as the MSE implicitly does, won't work in this case. Take, for example, these three pictures of a stained glass window in the Church of Saint John the Theologian, taken from the Library of Congress Prokudin-Gorskii Collection (*http://www.loc.gov/pictures/item/prk2000000263/*) (see Figure 7-5):

```
from skimage import io
stained_glass = io.imread('data/00998v.jpg') / 255  # use float image in [0, 1]
fig, ax = plt.subplots(figsize=(4.8, 7))
ax.imshow(stained_glass)
ax.axis('off');
```

Figure 7-5. A Prokudin-Gorskii plate: three photos of the same stained glass window, taken with three different filters

Take a look at St. John's robes: they look pitch black in one image, gray in another, and bright white in the third! This would result in a terrible MSE score, even with perfect alignment.

Let's see what we can do with this. We start by splitting the plate into its component channels:

```
nrows = stained_glass.shape[0]
step = nrows // 3
channels = (stained_glass[:step],
            stained_glass[step:2*step],
            stained_glass[2*step:3*step])
channel_names = ['blue', 'green', 'red']
fig, axes = plt.subplots(1, 3)
for ax, image, name in zip(axes, channels, channel_names):
    ax.imshow(image)
    ax.axis('off')
    ax.set_title(name)
```

First, we overlay all three images to verify that the alignment indeed needs to be fine-tuned between the three channels:

```
blue, green, red = channels
original = np.dstack((red, green, blue))
fig, ax = plt.subplots(figsize=(4.8, 4.8), tight_layout=True)
ax.imshow(original)
ax.axis('off');
```

You can see by the color "halos" around objects in the image that the colors are close to alignment, but not quite. Let's try to align them in the same way that we aligned the astronaut image above, using the MSE. We use one color channel, green, as the reference image, and align the blue and red channels to that.

```
print('*** Aligning blue to green ***')
tf = align(green, blue)
cblue = transform.warp(blue, tf, order=3)

print('** Aligning red to green ***')
tf = align(green, red)
cred = transform.warp(red, tf, order=3)

corrected = np.dstack((cred, green, cblue))
f, (ax0, ax1) = plt.subplots(1, 2)
ax0.imshow(original)
ax0.set_title('Original')
ax1.imshow(corrected)
ax1.set_title('Corrected')
```

```
for ax in (ax0, ax1):
    ax.axis('off')

*** Aligning blue to green ***
Level: 1, Angle: -0.0474, Offset: (-0.867, 15.4), Cost: 0.0499
** Aligning red to green ***
Level: 1, Angle: 0.0339, Offset: (-0.269, -8.88), Cost: 0.0311
```

The alignment is a little bit better than with the raw images (Figure 7-6), because the red and the green channels are correctly aligned, probably thanks to the giant yellow patch of sky. However, the blue channel is still off, because the bright spots of blue don't coincide with the green channel. That means that the MSE will be lower when the channels are *mis*-aligned so that blue patches overlap with some bright green spots.

Figure 7-6. MSE-based alignment reduces but does not eliminate the color halos

We turn instead to a measure called *normalized mutual information* (NMI), which measures correlations between the different brightness bands of the different images. When the images are perfectly aligned, any object of uniform color will create a large correlation between the shades of the different component channels, and a correspondingly large NMI value. In a sense, NMI measures how easy it would be to predict a pixel value of one image given the value of the corresponding pixel in the other. It was defined in the paper "An Overlap Invariant Entropy Measure of 3D Medical Image Alignment" (*http://bit.ly/2trbaFu*):[3]

$$I(X, Y) = \frac{H(X)+H(Y)}{H(X,Y)},$$

---

3 C. Studholme, D. L. G. Hill, and D. J. Hawkes, "An Overlap Invariant Entropy Measure of 3D Medical Image Alignment," Pattern Recognition 32, no. 1 (1999): 71–86.

where $H(X)$ is the *entropy* of X, and $H(X,Y)$ is the joint entropy of X and Y. The numerator describes the entropy of the two images, seen separately, and the denominator the total entropy if they are observed together. Values can vary between 1 (maximally aligned) and 2 (minimally aligned).[4] See Chapter 5.

In Python code, this becomes:

```
from scipy.stats import entropy

def normalized_mutual_information(A, B):
    """Compute the normalized mutual information.

    The normalized mutual information is given by:

              H(A) + H(B)
    Y(A, B) = -----------
                H(A, B)

    where H(X) is the entropy ``- sum(x log x) for x in X``.

    Parameters
    ----------
    A, B : ndarray
        Images to be registered.

    Returns
    -------
    nmi : float
        The normalized mutual information between the two arrays, computed at a
        granularity of 100 bins per axis (10,000 bins total).
    """
    hist, bin_edges = np.histogramdd([np.ravel(A), np.ravel(B)], bins=100)
    hist /= np.sum(hist)

    H_A = entropy(np.sum(hist, axis=0))
    H_B = entropy(np.sum(hist, axis=1))
    H_AB = entropy(np.ravel(hist))

    return (H_A + H_B) / H_AB
```

Now we define a *cost function* to optimize, as we defined `cost_mse` above:

```
def cost_nmi(param, reference_image, target_image):
    transformation = make_rigid_transform(param)
    transformed = transform.warp(target_image, transformation, order=3)
    return -normalized_mutual_information(reference_image, transformed)
```

---

4 A quick hand-wavy explanation is that entropy is calculated from the histogram of the quantity under consideration. If $X = Y$, then the joint histogram (X, Y) is diagonal, and that diagonal is the same as that of either X or Y. Thus, $H(X) = H(Y) = H(X, Y)$ and $I(X, Y) = 2$.

Finally, we use this with our basinhopping-optimizing aligner (Figure 7-7):

```
print('*** Aligning blue to green ***')
tf = align(green, blue, cost=cost_nmi)
cblue = transform.warp(blue, tf, order=3)

print('** Aligning red to green ***')
tf = align(green, red, cost=cost_nmi)
cred = transform.warp(red, tf, order=3)

corrected = np.dstack((cred, green, cblue))
fig, ax = plt.subplots(figsize=(4.8, 4.8), tight_layout=True)
ax.imshow(corrected)
ax.axis('off')

*** Aligning blue to green ***
Level: 1, Angle: 0.444, Offset: (6.07, 0.354), Cost: -1.08
** Aligning red to green ***
Level: 1, Angle: 0.000657, Offset: (-0.635, -7.67), Cost: -1.11

(-0.5, 393.5, 340.5, -0.5)
```

Figure 7-7. Prokudin-Gorskii channels aligned with normalized mutual information

What a glorious image! Realize that this artifact was created before color photography existed! Notice God's pearly white robes, John's white beard, and the white pages of the book held by Prochorus, his scribe—all of which were missing from the MSE-based alignment, but look wonderfully clear using NMI. Notice also the realistic gold of the candlesticks in the foreground.

We've illustrated the two key concepts in function optimization in this chapter: understanding local minima and how to avoid them, and choosing the right function to optimize to achieve a particular objective. Solving these allows you to apply optimization to a wide array of scientific problems!

## CHAPTER 8

Big Data in Little Laptop with Toolz

GRACIE: A knife? The guy's twelve feet tall!
JACK: Seven. Hey, don't worry, I think I can handle him.
— Jack Burton, *Big Trouble in Little China*

Streaming is not a SciPy feature per se, but rather an approach that allows us to efficiently process large datasets, like those often seen in science. The Python language contains some useful primitives for streaming data processing, and these can be combined with Matt Rocklin's Toolz library to generate elegant, concise code that is extremely memory-efficient. In this chapter, we will show you how to apply these streaming concepts to enable you to handle much larger datasets than can fit in your computer's RAM.

You have probably already done some streaming, perhaps without thinking about it in these terms. The simplest form is probably iterating through lines in a files, processing each line without ever reading the entire file into memory. For example, a loop like this to calculate the mean of each row and sum them:

```
import numpy as np
with open('data/expr.tsv') as f:
    sum_of_means = 0
    for line in f:
        sum_of_means += np.mean(np.fromstring(line, dtype=int, sep='\t'))
print(sum_of_means)
```

```
1463.0
```

This strategy works really well for cases where your problem can be neatly solved with by-row processing. But things can quickly get out of hand when your code becomes more sophisticated.

In streaming programs, a function processes *some* of the input data, returns the processed chunk, then, while downstream functions are dealing with that chunk, the function receives a bit more, and so on. All these things are going on at the same time! How can one keep them straight?

We too found this difficult, until we discovered the Toolz library. Its constructs make streaming programs so elegant to write that it was impossible to contemplate writing this book without including a chapter about it.

Let us clarify what we mean by "streaming" and why you might want to do it. Suppose you have some data in a text file, and you want to compute the column-wise average of $\log(x + 1)$ of the values. The common way to do this would be to use NumPy to load the values, compute the log function for all values in the full matrix, and then take the mean over the first axis:

```
import numpy as np
expr = np.loadtxt('data/expr.tsv')
logexpr = np.log(expr + 1)
np.mean(logexpr, axis=0)
```

```
array([ 3.11797294,  2.48682887,  2.19580049,  2.36001866,  2.70124539,
        2.64721531,  2.43704834,  3.28539133,  2.05363724,  2.37151577,
        3.85450782,  3.9488385 ,  2.46680157,  2.36334423,  3.18381635,
        2.64438124,  2.62966516,  2.84790568,  2.61691451,  4.12513405])
```

This works, and it follows a reassuringly familiar input-output model of computation. But it's a pretty inefficient way to go about it! We load the full matrix into memory (1), then make a copy with 1 added to each value (2), then make another copy to compute the log (3), before finally passing it on to np.mean. That's three instances of the data array, to perform an operation that doesn't require keeping even *one* instance in memory. For any kind of "big data" operation, this approach won't work.

Python's creators knew this, and created the yield keyword, which enables a function to process just one "sip" of the data, pass the result on to the next process, and *let the chain of processing complete* for that one piece of data before moving on to the next one. "Yield" is a rather nice name for it: the function *yields* control to the next function, waiting to resume processing the data until all the downstream steps have processed that data point.

Streaming with yield

The flow of control described above can be rather hard to follow. An awesome feature of Python is that it abstracts this complexity away, allowing you to focus on the analysis functionality. Here's one way to think about it: for every processing function that would normally take a list (a collection of data) and transform that list, you can rewrite that function as taking a *stream* and *yielding* the result of every element of that stream.

Here's an example where we take the log of each element in a list, using either a standard data-copying method or a streaming method:

```
def log_all_standard(input):
    output = []
    for elem in input:
        output.append(np.log(elem))
    return output

def log_all_streaming(input_stream):
    for elem in input_stream:
        yield np.log(elem)
```

Let's check that we get the same result with both methods:

```
# We set the random seed so we will get consistent results
np.random.seed(seed=7)
# Set print options to show only 3 significant digits
np.set_printoptions(precision=3, suppress=True)

arr = np.random.rand(1000) + 0.5
result_batch = sum(log_all_standard(arr))
print('Batch result: ', result_batch)
result_stream = sum(log_all_streaming(arr))
print('Stream result: ', result_stream)

Batch result:  -48.2409194561
Stream result:  -48.2409194561
```

The advantage of the streaming approach is that elements of a stream aren't processed until they're needed, whether it's for computing a running sum, or for writing out to disk, or something else. This can conserve a lot of memory when you have many input items, or when each item is very big. (Or both!) This quote from one of Matt's blog posts (*http://bit.ly/2trkKZ6*) very succinctly summarizes the utility of streaming data analysis:

> In my brief experience people rarely take this [streaming] route. They use single-threaded in-memory Python until it breaks, and then seek out Big Data Infrastructure like Hadoop/Spark at relatively high productivity overhead.

Indeed, this describes our computational careers perfectly. But the intermediate approach can get you a *lot* farther than you think. In some cases, it can get you there even faster than the supercomputing approach, by eliminating the overhead of multicore communication and random access to databases. (For example, see the blog post "Bigger data; same laptop" (*http://bit.ly/2trD0BL*) by Frank McSherry, where he processes a 128 billion edge graph on his laptop *faster* than using a graph database on a supercomputer.)

To clarify the flow of control when using streaming-style functions, it's useful to make *verbose* versions of the functions, which print out a message with each operation.

```
import numpy as np

def tsv_line_to_array(line):
    lst = [float(elem) for elem in line.rstrip().split('\t')]
    return np.array(lst)

def readtsv(filename):
    print('starting readtsv')
    with open(filename) as fin:
        for i, line in enumerate(fin):
            print(f'reading line {i}')
            yield tsv_line_to_array(line)
    print('finished readtsv')

def add1(arrays_iter):
    print('starting adding 1')
    for i, arr in enumerate(arrays_iter):
        print(f'adding 1 to line {i}')
        yield arr + 1
    print('finished adding 1')

def log(arrays_iter):
    print('starting log')
    for i, arr in enumerate(arrays_iter):
        print(f'taking log of array {i}')
        yield np.log(arr)
    print('finished log')

def running_mean(arrays_iter):
    print('starting running mean')
    for i, arr in enumerate(arrays_iter):
        if i == 0:
            mean = arr
        mean += (arr - mean) / (i + 1)
        print(f'adding line {i} to the running mean')
    print('returning mean')
    return mean
```

Let's see it in action for a small sample file:

```
fin = 'data/expr.tsv'
print('Creating lines iterator')
lines = readtsv(fin)
print('Creating loglines iterator')
loglines = log(add1(lines))
print('Computing mean')
mean = running_mean(loglines)
print(f'the mean log-row is: {mean}')

Creating lines iterator
Creating loglines iterator
Computing mean
starting running mean
```

```
starting log
starting adding 1
starting readtsv
reading line 0
adding 1 to line 0
taking log of array 0
adding line 0 to the running mean
reading line 1
adding 1 to line 1
taking log of array 1
adding line 1 to the running mean
reading line 2
adding 1 to line 2
taking log of array 2
adding line 2 to the running mean
reading line 3
adding 1 to line 3
taking log of array 3
adding line 3 to the running mean
reading line 4
adding 1 to line 4
taking log of array 4
adding line 4 to the running mean
finished readtsv
finished adding 1
finished log
returning mean
the mean log-row is: [ 3.118  2.487  2.196  2.36   2.701  2.647  2.437  3.285
                       2.054  2.372
   3.855  3.949  2.467  2.363  3.184  2.644  2.63   2.848  2.617  4.125]
```

Note:

- None of the computation is run when creating the lines and loglines iterators. This is because iterators are *lazy*, meaning they are not evaluated (or *consumed*) until a result is needed.
- When the computation is finally triggered, by the call to running_mean, it jumps back and forth between all the functions, as various computations are performed on each line, before moving on to the next line.

Introducing the Toolz Streaming Library

In this chapter's code example, contributed by Matt Rocklin, we create a Markov model from an entire fly genome in under five minutes on a laptop, using just a few lines of code. (We have slightly edited it for easier downstream processing.) Matt's example uses a human genome, but apparently our laptops weren't quite so fast, so we're going to use a fly genome instead (it's about 1/20 the size). Over the course of the chapter we'll actually augment it a little bit to start from compressed data (who wants to keep an uncompressed dataset on their hard drive?). This modification is almost *trivial*, which speaks to the elegance of his example.

```python
import toolz as tz
from toolz import curried as c
from glob import glob
import itertools as it

LDICT = dict(zip('ACGTacgt', range(8)))
PDICT = {(a, b): (LDICT[a], LDICT[b])
         for a, b in it.product(LDICT, LDICT)}

def is_sequence(line):
    return not line.startswith('>')

def is_nucleotide(letter):
    return letter in LDICT  # ignore 'N'

@tz.curry
def increment_model(model, index):
    model[index] += 1


def genome(file_pattern):
    """Stream a genome, letter by letter, from a list of FASTA filenames."""
    return tz.pipe(file_pattern, glob, sorted,  # Filenames
                c.map(open),  # lines
                # concatenate lines from all files:
                tz.concat,
                # drop header from each sequence
                c.filter(is_sequence),
                # concatenate characters from all lines
                tz.concat,
                # discard newlines and 'N'
                c.filter(is_nucleotide))


def markov(seq):
    """Get a 1st-order Markov model from a sequence of nucleotides."""
    model = np.zeros((8, 8))
    tz.last(tz.pipe(seq,
                c.sliding_window(2),        # each successive tuple
                c.map(PDICT.__getitem__),   # location in matrix of tuple
                c.map(increment_model(model))))  # increment matrix
    # convert counts to transition probability matrix
    model /= np.sum(model, axis=1)[:, np.newaxis]
    return model
```

We can then do the following to obtain a Markov model of repetitive sequences in the fruit fly genome:

```python
%%timeit -r 1 -n 1
dm = 'data/dm6.fa'
```

```
model = tz.pipe(dm, genome, c.take(10**7), markov)
# we use `take` to just run on the first 10 million bases, to speed things up.
# the take step can just be removed if you have ~5-10 mins to wait.

1 loop, average of 1: 24.3 s +- 0 ns per loop (using standard deviation)
```

There's a *lot* going on in that example, so we are going to unpack it little by little. We'll actually run the example at the end of the chapter.

The first thing to note is how many functions come from the Toolz library (*http://toolz.readthedocs.org/en/latest/*). For example, from Toolz we've used pipe, slid ing_window, frequencies, and a curried version of map (more on this later). That's because Toolz is written specifically to take advantage of Python's iterators and easily manipulate streams.

Let's start with pipe. This function is simply syntactic sugar to make nested function calls easier to read. This is important because that pattern becomes increasingly common when you're dealing with iterators.

As a simple example, let's rewrite our running mean using pipe:

```
import toolz as tz
filename = 'data/expr.tsv'
mean = tz.pipe(filename, readtsv, add1, log, running_mean)

# This is equivalent to nesting the functions like this:
# running_mean(log(add1(readtsv(filename))))

starting running mean
starting log
starting adding 1
starting readtsv
reading line 0
adding 1 to line 0
taking log of array 0
adding line 0 to the running mean
reading line 1
adding 1 to line 1
taking log of array 1
adding line 1 to the running mean
reading line 2
adding 1 to line 2
taking log of array 2
adding line 2 to the running mean
reading line 3
adding 1 to line 3
taking log of array 3
adding line 3 to the running mean
reading line 4
adding 1 to line 4
taking log of array 4
adding line 4 to the running mean
```

```
finished readtsv
finished adding 1
finished log
returning mean
```

What was originally multiple lines, or an unwieldy mess of parentheses, is now a clean description of the sequential transformations of the input data. Much easier to understand!

This strategy also has an advantage over the original NumPy implementation: if we scale our data to millions or billions of rows, our computer might struggle to hold all the data in memory. In contrast, here we are only loading lines from disk one at a time, and maintaining only a single line's worth of data.

k-mer Counting and Error Correction

You might want to review Chapters 1 and 2 for information about DNA and genomics. Briefly, your genetic information, the blueprint for making *you*, is encoded as a sequence of chemical *bases* in your *genome*. These are really, really tiny, so you can't just look in a microscope and read them. You also can't read a long string of them: errors accumulate and the readout becomes unusable. (New technology is changing this, but here we will focus on short-read sequencing data, the most common today.) Luckily, every one of your cells has an identical copy of your genome, so what we can do is shred those copies into tiny segments (about 100 bases long), and then assemble those like an enormous puzzle of 30 million pieces.

Before performing assembly, it is vital to perform read correction. During DNA sequencing some bases are incorrectly read out, and must be fixed, or they will mess up the assembly. (Imagine having puzzle pieces with the wrong shape.)

One correction strategy is to find similar reads in your dataset and fix the error by grabbing the correct information from those reads. Or alternatively, you may choose to completely discard those reads containing errors.

However, this is a very inefficient way to do it, because finding similar reads means you would compare each read to every other read. This takes N^2 operations, or 9×10^{14} for a 30 million read dataset! (And these are not cheap operations.)

There is another way. Pavel Pevzner and others (*http://www.pnas.org/content/98/17/9748.full*) realized that reads could be broken down into smaller, overlapping *k-mers*, substrings of length *k*, which can then be stored in a hash table (a dictionary in Python). This has tons of advantages, but the main one is that instead of computing on the total number of reads, which can be arbitrarily large, we can compute on the total number of k-mers, which can only be as large as the genome itself—usually one to two orders of magnitude smaller than the reads.

If we choose a value for *k* that is large enough to ensure any k-mer appears only once in the genome, the number of times a k-mer appears is exactly the number of reads that originate from that part of the genome. This is called the *coverage* of that region.

If a read has an error in it, there is a high probability that the k-mers overlapping the error will be unique or close to unique in the genome. Think of the equivalent in English: if you were to take reads from Shakespeare, and one read was "to be or nob to be," the 6-mer "nob to" will appear rarely or not at all, whereas "not to" will be very frequent.

This is the basis for k-mer error correction: split the reads into k-mers, count the occurrence of each k-mer, and use some logic to replace rare k-mers in reads with similar common ones. (Or, alternatively, discard reads with erroneous k-mers. This is possible because reads are so abundant that we can afford to toss out erroneous data.)

This is also an example in which streaming is *essential*. As mentioned before, the number of reads can be enormous, so we don't want to store them in memory.

DNA sequence data is commonly represented in FASTA format. This is a plain-text format, consisting of one or many DNA sequences per file, each with a name and the actual sequence.

A sample FASTA file:

```
> sequence_name1
TCAATCTCTTTTATATTAGATCTCGTTAAAGTAAAATTTTGGTTTGTGTTAAAGTACAAG
GGGTACCTATGACCACGGAACCAACAAAGTGCCTAAATAGGACATCAAGTAACTAGCGGT
ACGT

> sequence_name2
ATGTCCCAGGCGTTCCTTTTGCATTTGCTTCGCATTAACAGAATATCCAGCGTACTTAGG
ATTGTCGACCTGTCTTGTCGTACGTGGCCGCAACACCAGGTATAGTGCCAATACAAGTCA
GACTAAAACTGGTTC
```

Now we have the required information to convert a stream of lines from a FASTA file to a count of k-mers:

- Filter lines so that only sequence lines are used
- For each sequence line, produce a stream of k-mers
- Add each k-mer to a dictionary counter

Here's how you would do this in pure Python, using nothing but built-ins:

```python
def is_sequence(line):
    line = line.rstrip()  # remove '\n' at end of line
    return len(line) > 0 and not line.startswith('>')

def reads_to_kmers(reads_iter, k=7):
    for read in reads_iter:
        for start in range(0, len(read) - k):
```

```
            yield read[start : start + k]  # note yield, so this is a generator

def kmer_counter(kmer_iter):
    counts = {}
    for kmer in kmer_iter:
        if kmer not in counts:
            counts[kmer] = 0
        counts[kmer] += 1
    return counts

with open('data/sample.fasta') as fin:
    reads = filter(is_sequence, fin)
    kmers = reads_to_kmers(reads)
    counts = kmer_counter(kmers)
```

This totally works and is streaming, so reads are loaded from disk one at a time and
piped through the k-mer converter and to the k-mer counter. We can then plot a histogram of the counts, and confirm that there are indeed two well-separated popula-
tions of correct and erroneous k-mers:

```
# Make plots appear inline, set custom plotting style
%matplotlib inline
import matplotlib.pyplot as plt
plt.style.use('style/elegant.mplstyle')

def integer_histogram(counts, normed=True, xlim=[], ylim=[],
                      *args, **kwargs):
    hist = np.bincount(counts)
    if normed:
        hist = hist / np.sum(hist)
    fig, ax = plt.subplots()
    ax.plot(np.arange(hist.size), hist, *args, **kwargs)
    ax.set_xlabel('counts')
    ax.set_ylabel('frequency')
    ax.set_xlim(*xlim)
    ax.set_ylim(*ylim)

counts_arr = np.fromiter(counts.values(), dtype=int, count=len(counts))
integer_histogram(counts_arr, xlim=(-1, 250))
```

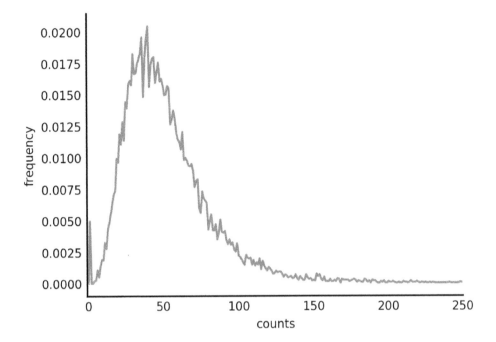

Notice the nice distribution of k-mer frequencies, along with a big bump of k-mers (at the left of the plot) that appear only once. Such low-frequency k-mers are likely to be errors.

But, with the preceding code, we are actually doing a bit too much work. A lot of the functionality we wrote in for loops and yields is actually *stream manipulation*: transforming a stream of data into a different kind of data, and accumulating it at the end. Toolz has a lot of stream-manipulation primitives that make it easy to write the above in just one function call; and, once you know the names of the transforming functions, it also becomes easier to visualize what is happening to your data stream at each point.

For example, the *sliding window* function is exactly what we need to make k-mers:

```
print(tz.sliding_window.__doc__)

A sequence of overlapping subsequences

    >>> list(sliding_window(2, [1, 2, 3, 4]))
    [(1, 2), (2, 3), (3, 4)]

    This function creates a sliding window suitable for transformations like
    sliding means / smoothing

    >>> mean = lambda seq: float(sum(seq)) / len(seq)
```

```
>>> list(map(mean, sliding_window(2, [1, 2, 3, 4])))
[1.5, 2.5, 3.5]
```

Additionally, the *frequencies* function counts the appearance of individual items in a data stream. Together with pipe, we can now count k-mers in a single function call:

```
from toolz import curried as c

k = 7
counts = tz.pipe('data/sample.fasta', open,
                 c.filter(is_sequence),
                 c.map(str.rstrip),
                 c.map(c.sliding_window(k)),
                 tz.concat, c.map(''.join),
                 tz.frequencies)
```

But just a minute: what are all those c.function calls from toolz.curried?

Currying: The Spice of Streaming

Earlier, we briefly used a *curried* version of the map function, which applies a given function to each element in a sequence. Now that we've mixed a few more curried calls in there, it's time to share with you what it means! Currying is not named after the spice blend (though it does spice up your code). It is named for Haskell Curry, the mathematician who invented the concept. Haskell Curry is also the namesake of the Haskell programming language—in which *all* functions are curried!

"Currying" means *partially* evaluating a function and returning another, "smaller" function. Normally in Python if you don't give a function all of its required arguments, it will throw a fit. In contrast, a curried function can just take *some* of those arguments. If the curried function doesn't get enough arguments, it returns a new function that takes the leftover arguments. Once that second function is called with the remaining arguments, it can perform the original task. Another term for currying is *partial evaluation*. In functional programming, currying is a way to produce a function that can wait for the rest of the arguments to show up later.

So, while the function call map(np.log, numbers_list) applies the np.log function to all of the numbers in numbers_list (returning a sequence of the logged numbers), the call toolz.curried.map(np.log) returns a *function* that takes in a sequence of numbers and returns a sequence of logged numbers.

It turns out that having a function that already knows about some of the arguments is perfect for streaming! We've seen a hint of how powerful currying and pipes can be together in the above code snippet.

But currying can be a bit of a mind-bend when you first start, so we'll try it with some simple examples to demonstrate how it works. Let's start by writing a simple, non-curried function:

```
def add(a, b):
    return a + b

add(2, 5)
```

```
7
```

Now we write a similar function which we curry manually:

```
def add_curried(a, b=None):
    if b is None:
        # second argument not given, so make a function and return it
        def add_partial(b):
            return add(a, b)
        return add_partial
    else:
        # Both values were given, so we can just return a value
        return add(a, b)
```

Now let's try out a curried function to make sure it does what we expect:

```
add_curried(2, 5)
```

```
7
```

Okay, it acts like a normal function when given both variables. Now let's leave out the second variable:

```
add_curried(2)
```

```
<function __main__.add_curried.<locals>.add_partial>
```

As we expected, it returned a function. Now let's use that function:

```
add2 = add_curried(2)
add2(5)
```

```
7
```

Now, that worked, but `add_curried` was a hard function to read. Future us will probably have trouble remembering how we wrote that code. Luckily, Toolz has the, well, tools to help us out.

```
import toolz as tz

@tz.curry  # Use curry as a decorator
def add(x, y):
    return x + y

add_partial = add(2)
add_partial(5)
```

```
7
```

To summarize what we did, add is now a curried function, so it can take one of the arguments and returns another function, add_partial, which "remembers" that argument.

In fact, all of the Toolz functions are also available as curried functions in the toolz.curried namespace. Toolz also includes curried versions of some handy higher-order Python functions like map, filter, and reduce. We will import the curried namespace as c so our code doesn't get too cluttered. So, for example, the curried version of map will be c.map. Note that the curried functions (e.g., c.map) are different from the @curry decorators, which are used to *create* curried functions.

```
from toolz import curried as c
c.map

<class 'map'>
```

As a reminder, map is a built-in function. From the docs (*https://docs.python.org/3.4/library/functions.html#map*):

map(function, iterable, ...) Return an iterator that applies function to every item of iterable, yielding the results.

A curried version of map is particularly handy when working in a Toolz pipe. You can just pass a function to c.map and then stream in the iterator later using tz.pipe. Take another look at our function for reading in the genome to see how this works in practice.

```
def genome(file_pattern):
    """Stream a genome, letter by letter, from a list of FASTA filenames."""
    return tz.pipe(file_pattern, glob, sorted,  # Filenames
                   c.map(open),  # lines
                   # concatenate lines from all files:
                   tz.concat,
                   # drop header from each sequence
                   c.filter(is_sequence),
                   # concatenate characters from all lines
                   tz.concat,
                   # discard newlines and 'N'
                   c.filter(is_nucleotide))
```

Back to Counting k-mers

Okay, so now we've got our heads around currying, let's get back to our k-mer counting code. Here's that code again that used those curried functions:

```
from toolz import curried as c

k = 7
counts = tz.pipe('data/sample.fasta', open,
                 c.filter(is_sequence),
```

```
            c.map(str.rstrip),
            c.map(c.sliding_window(k)),
            tz.concat, c.map(''.join),
            tz.frequencies)
```

We can now observe the frequency of different k-mers:

```
counts = np.fromiter(counts.values(), dtype=int, count=len(counts))
integer_histogram(counts, xlim=(-1, 250), lw=2)
```

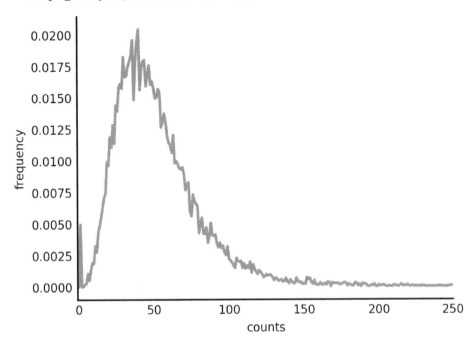

Tips for Working with Streams

- Convert "list of list" to "long list" with tz.concat.
- Don't get caught out:
  - Iterators get consumed. So if you make a generator object and do some processing on it, and then a later step fails, you need to re-create the generator. The original is already gone.
  - Iterators are lazy. You need to force evaluation sometimes.
- When you have lots of functions in a pipe, it's sometimes hard to figure out where things went wrong. Take a small stream and add functions to your pipe one by one from the first/leftmost until you find the broken one. You can also insert map(do(print)) (map and do are from toolz.curried) at any point in a stream to print each element while it streams through.

Exercise: PCA of Streaming Data

The scikit-learn library has an `IncrementalPCA` class, which allows you to run principal components analysis on a dataset without loading the full dataset into memory. But you need to chunk your data yourself, which makes the code a bit awkward to use. Make a function that can take a stream of data samples and perform PCA. Then, use the function to compute the PCA of the `iris` machine learning dataset, which is in *data/iris.csv*. (You can also access it from the `datasets` module of scikit-learn, using `datasets.load_iris()`.) Optionally, you can color the points with the species number, found in *data/iris-target.csv*.

The `IncrementalPCA` class is in `sklearn.decomposition`, and requires a *batch size* greater than 1 to train the model. Look at the `toolz.curried.partition` function for how to create a stream of batches from a stream of data points.

Markov Model from a Full Genome

Back to our original code example. What is a Markov model, and why is it useful?

In general, a Markov model assumes that the probability of the system moving to a given state is only dependent on the state that it was in just previously. For example, if it is sunny right now, there is a high probability that it will be sunny tomorrow. The fact that it was raining yesterday is irrelevant. In this theory, all the information required to predict the future is encoded in the current state of things. The past is irrelevant. This assumption is useful for simplifying otherwise intractable problems, and often gives good results. Markov models are behind much of the signal processing in mobile phone and satellite communications, for example.

In the context of genomics, as we will see, different functional regions of a genome have different *transition probabilities* between similar states. Observing these in a new genome, we can predict something about the function of those regions. Going back to the weather analogy, the probability of going from a sunny day to a rainy day is very different depending on whether you are in Los Angeles or London. Therefore, if I give you a string of (sunny, sunny, sunny, rainy, sunny, ...) days, you can predict whether it came from Los Angeles or London, assuming you have a previously trained model.

In this chapter, we'll cover just the model building, for now.

You can download the *Drosophila melanogaster* (fruit fly) genome file, *dm6.fa.gz* (*http://hgdownload.cse.ucsc.edu/goldenPath/dm6/bigZips/*). You will need to unzip it using `gzip -d dm6.fa.gz`.

In the genome data, the genetic sequence, which consists of the letters A, C, G, and T, is encoded as belonging to *repetitive elements*, a specific class of DNA, by whether it is in lowercase (repetitive) or uppercase (nonrepetitive). We can use this information when we build the Markov model.

We want to encode the Markov model as a NumPy array, so we will make dictionaries to index from letters to indices in [0, 7] (LDICT for "letters dictionary"), and from pairs of letters to 2D indices in ([0, 7], [0, 7]) (PDICT or "pairs dictionary"):

```
import itertools as it

LDICT = dict(zip('ACGTacgt', range(8)))
PDICT = {(a, b): (LDICT[a], LDICT[b])
         for a, b in it.product(LDICT, LDICT)}
```

We also want to filter out nonsequence data: the sequence names, which are in lines starting with >, and unknown sequence, which is labeled as N, so we will make functions to filter on:

```
def is_sequence(line):
    return not line.startswith('>')

def is_nucleotide(letter):
    return letter in LDICT  # ignore 'N'
```

Finally, whenever we get a new nucleotide pair, say, (A, T), we want to increment our Markov model (our NumPy matrix) at the corresponding position. We make a curried function to do so:

```
import toolz as tz

@tz.curry
def increment_model(model, index):
    model[index] += 1
```

We can now combine these elements to stream a genome into our NumPy matrix. Note that, if seq below is a stream, we never need to store the whole genome, or even a big chunk of the genome, in memory!

```
from toolz import curried as c

def markov(seq):
    """Get a 1st-order Markov model from a sequence of nucleotides."""
    model = np.zeros((8, 8))
    tz.last(tz.pipe(seq,
                    c.sliding_window(2),      # each successive tuple
                    c.map(PDICT.__getitem__), # location in matrix of tuple
                    c.map(increment_model(model)))) # increment matrix
    # convert counts to transition probability matrix
    model /= np.sum(model, axis=1)[:, np.newaxis]
    return model
```

Now we simply need to produce that genome stream, and make our Markov model:

```
from glob import glob

def genome(file_pattern):
    """Stream a genome, letter by letter, from a list of FASTA filenames."""
    return tz.pipe(file_pattern, glob, sorted,  # Filenames
                   c.map(open),  # lines
                   # concatenate lines from all files:
                   tz.concat,
                   # drop header from each sequence
                   c.filter(is_sequence),
                   # concatenate characters from all lines
                   tz.concat,
                   # discard newlines and 'N'
                   c.filter(is_nucleotide))
```

Let's try it out on the Drosophila (fruit fly) genome:

```
# Download dm6.fa.gz from ftp://hgdownload.cse.ucsc.edu/goldenPath/dm6/bigZips/
# Unzip before using: gzip -d dm6.fa.gz
dm = 'data/dm6.fa'
model = tz.pipe(dm, genome, c.take(10**7), markov)
# we use `take` to just run on the first 10 million bases, to speed things up.
# the take step can just be removed if you have ~5-10 mins to wait.
```

Let's look at the resulting matrix:

```
print('    ', '    '.join('ACGTacgt'), '\n')
print(model)

        A      C      G      T      a      c      g      t
[[ 0.348  0.182  0.194  0.275  0.     0.     0.     0.    ]
 [ 0.322  0.224  0.198  0.254  0.     0.     0.     0.    ]
 [ 0.262  0.272  0.226  0.239  0.     0.     0.     0.    ]
 [ 0.209  0.199  0.245  0.347  0.     0.     0.     0.    ]
 [ 0.003  0.003  0.003  0.003  0.349  0.178  0.166  0.296]
 [ 0.002  0.002  0.003  0.003  0.376  0.195  0.152  0.267]
 [ 0.002  0.003  0.003  0.002  0.281  0.231  0.194  0.282]
 [ 0.002  0.002  0.003  0.003  0.242  0.169  0.227  0.351]]
```

It's probably clearer to look at the result as an image (Figure 8-1):

```
def plot_model(model, labels, figure=None):
    fig = figure or plt.figure()
    ax = fig.add_axes([0.1, 0.1, 0.8, 0.8])
    im = ax.imshow(model, cmap='magma');
    axcolor = fig.add_axes([0.91, 0.1, 0.02, 0.8])
    plt.colorbar(im, cax=axcolor)
    for axis in [ax.xaxis, ax.yaxis]:
        axis.set_ticks(range(8))
        axis.set_ticks_position('none')
        axis.set_ticklabels(labels)
    return ax
```

```
plot_model(model, labels='ACGTacgt');
```

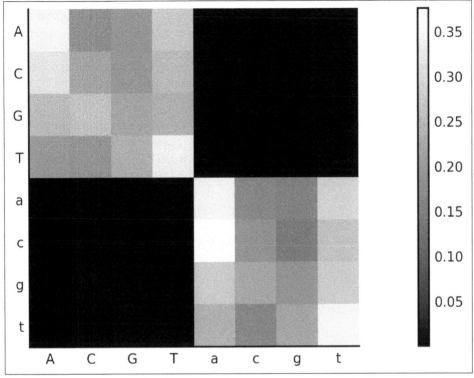

Figure 8-1. Transition probability matrix for genetic sequence in the Drosophila mela-nogaster genome

Notice how the C–A and G–C transitions are different between the repeat and non-repeat parts of the genome. This information can be used to classify a previously unseen DNA sequence.

Exercise: Online Unzip

Add a step to the start of the pipe to unzip the data so you don't have to keep a decompressed version on your hard drive. The Drosophila genome, for example, takes less than a third of the space on disk when compressed with gzip. And yes, unzipping can be streamed, too!

> The gzip package, part of Python's standard library, allows you to open *.gz* files as if they were normal files.

We hope to have shown you at least a hint that streaming in Python can be easy when you use a few abstractions, like the ones Toolz provides.

Streaming can make you more productive, because big data takes linearly longer than small data. In batch analysis, big data can take forever to run, because the operating system has to keep transferring data from RAM to the hard disk and back. Or, Python might refuse altogether and simply show a MemoryError! This means that, for many analyses, you don't need a bigger machine to analyze bigger datasets. And, if your tests pass on small data, they'll pass on big data, too!

Our take-home message from this chapter is this: when writing an algorithm or analysis, think about whether you can do it streaming. If you can, just do it from the beginning. Your future self will thank you. Doing it later is harder, and results in things like Figure 8-2.

Figure 8-2. TODOs in history (http://bit.ly/2sXPg9u) (comic by Manu Cornet, used with permission)

Epilogue

Quality means doing it right when no one is looking.
—Henry Ford

Our main goal with this book was to promote elegant uses of the NumPy and SciPy libraries. While teaching you how to do effective scientific analysis with SciPy, we hope to have inspired in you the feeling that quality code is something worth striving for.

Where to Next?

Now that you know enough SciPy to analyze whatever data gets thrown your way, how do you move forward? We said when we started that we couldn't hope to cover all there is to know about the library and all its offshoots. Before we part ways, we want to point you to the many resources available to help you.

Mailing Lists

We mentioned in the preface that SciPy is a community. A great way to continue learning is to subscribe to the main mailing lists for NumPy, SciPy, pandas, Matplotlib, scikit-image, and other libraries you might be interested in, and read them regularly.

And when you do get stuck in your own work, don't be afraid to seek help there! We are a friendly bunch! The *main* requirement when seeking help is to show that you've tried a bit of problem solving yourself, and to provide others with a minimal script and enough sample data to demonstrate your problem and how you've tried to fix it.

- **No:** "I need to generate a big array of random Gaussians. Can someone help?"
- **No:** "I have this huge library at *https://github.com/ron_obvious*. If you look in the statistics library, there's a part that really needs random Gaussians. Can someone take a look???"

- **Yes:** "I've been trying to generate a big list of random Gaussians like so: `gauss = [np.random.randn()] * 10**5`. But when I compute `np.mean(gauss)`, it's hardly ever as close to 0 as I expect. What am I doing wrong? The full script is attached below."

GitHub

We also talked in the preface about GitHub. All of the code that we discussed lives on GitHub:

- NumPy (*https://github.com/numpy/numpy*)
- SciPy (*https://github.com/scipy/scipy*)

and others. When something isn't working as you expect, it could be a bug. If, after some investigation, you are convinced that you have indeed uncovered a bug, you should go to the "issues" tab of the relevant GitHub repository and create a new issue. This will ensure that the bug is on the radar of the developers of that library, and that it will (hopefully) be fixed in the next version. By the way, this advice also applies to "bugs" in the documentation: if something in a library's documentation isn't clear to you, file an issue!

Even better than filing an issue is *submitting a pull request*. A pull request improving a library's documentation is a great way to dip your toes in open source! We won't cover the process here, but there are many books and resources out there to help you:

- Anthony Scopatz and Katy Huff's *Effective Computation in Physics* (O'Reilly, 2015) covers Git and GitHub, among many other topics in scientific computation.
- *Introducing GitHub* (O'Reilly, 2015), by Peter Bell and Brent Beer, covers GitHub in more detail.
- Software Carpentry (*https://software-carpentry.org*) has Git lessons, and offers free workshops around the world throughout the year.
- Based partly on those lessons, one of your authors has created a complete tutorial on Git and GitHub pull requests, "Open Source Science with Git and GitHub" (*http://jni.github.io/git-tutorial/*).
- Finally, many open source projects on GitHub have a "CONTRIBUTING" (*http://bit.ly/2uFYZo5*) file, which contains a set of guidelines for contributing code to the project.

So, you are not starved for help on this topic!

We encourage you to contribute to the SciPy ecosystem as often as you can, not only because you will help make these libraries better for all, but also because it is one of the best ways to develop your coding abilities. With every pull request you submit, you will get feedback about your code, helping you to improve. You'll also become more familiar with the GitHub contributing process and etiquette, which are highly valuable skills in today's job market.

Conferences

In the same vein, we highly recommend attending a coding conference in this field. The SciPy conference, held every year in Austin, is fantastic, and probably your best bet if you enjoyed this book. There's also a European version, EuroSciPy, which changes host city every two years. Finally, the more general PyCon conference happens in the United States but also has offshoots around the world, such as PyCon-AU in Australia, which has a "Science and Data" miniconference the day before the main conference.

Whichever conference you choose, *stay for the sprints* at the end of the conference. A coding sprint is an intense session of team coding, and it is a fantastic opportunity to learn the process of contributing to open source, regardless of your skill level. It is how one of your authors (Juan) started in his open source journey.

Beyond SciPy

The SciPy library is written not just in Python, but also in highly optimized C and Fortran code that interfaces with Python. Together with NumPy and related libraries, it tries to cover most use cases that come up in scientific data analysis, and provides very fast functions for these. Sometimes, however, a scientific problem doesn't match at all with what's already in SciPy, and a pure Python solution is too slow to be useful. What to do then?

High Performance Python (O'Reilly, 2014), by Micha Gorelick and Ian Ozsvald, covers what you need to know in these situations: how to find where you *really* need performance, and the options available to get that performance. We highly recommend it.

Here, we want to briefly mention two of those options that are particularly relevant in the SciPy world.

First, Cython is a variant of Python that can be compiled to C, but then be imported into Python. Providing some type annotations to Python variables means the compiled C code can end up being a hundred or even a thousand times faster than comparable Python code. Cython is now an industry standard and is used in NumPy, SciPy, and many related libraries (such as scikit-image) to provide fast algorithms in array-based code. Kurt Smith has written the simply titled *Cython* (O'Reilly, 2015) to teach you the fundamentals of this language.

An often easier-to-use alternative to Cython is Numba, a just-in-time (JIT) compiler for array-based Python. JITs wait for a function to be executed once, at which point they can infer the types of all the function arguments and output, and compile the code into a highly efficient form for those specific types. In Numba code, you don't need to annotate types: Numba will infer them when a function is first called. Instead, you simply need to make sure that you only use basic types (integers, floats, etc.) and arrays rather than more complicated Python objects. In these cases, Numba can compile the Python code down to very efficient code and speed up computations by orders of magnitude.

Numba is still very young, but it is already very useful. Importantly, it shows what is possible by Python JITs, which are set to become more commonplace: Python 3.6 added features to make it easier to use new JITs (the Pyjion JIT is based on these). You can see some examples of Numba use, including how to combine it with SciPy, in Juan's blog (*https://ilovesymposia.com/tag/numba/*). And Numba, naturally, has its own very active and friendly mailing list.

Contributing to This Book

The source of this book is itself hosted on GitHub (*https://github.com/elegant-scipy/elegant-scipy*) (also at the *Elegant SciPy* website (*http://elegant-scipy.org*)). Just as if you were contributing to any other open source project, you can raise issues or submit pull requests if you find bugs or typos—and we would very much appreciate it if you did.

We used some of the best code we could find to illustrate the various parts of the SciPy and NumPy libraries. If you have a better example, please raise an issue in the repo. We would love to include it in future editions.

We are also on Twitter, at @elegantscipy (*https://twitter.com/elegantscipy*). Drop us a line if you want to chat about the book! The individual authors are @jnuneziglesias (*https://twitter.com/jnuneziglesias*), @stefanvdwalt (*https://twitter.com/stefanvdwalt*), and @hdashnow (*https://twitter.com/hdashnow*).

We particularly want to hear about it if you use any of the ideas or code in this book to advance your scientific research. That's the point of SciPy!

Until Next Time...

In the meantime, we hope you enjoyed this book and found it useful. If so, tell all your friends, and come say hi on the mailing lists, at a conference, on GitHub, and on Twitter. Thanks for reading, and here's to even more *Elegant SciPy*!

Exercise Solutions

Solution: Adding a Grid Overlay

This is the solution for "Exercise: Adding a Grid Overlay" on page 55.

We can use NumPy slicing to select the rows of the grid, set them to blue, and then select the columns and set them to blue as well (Figure A-1):

```
def overlay_grid(image, spacing=128):
    """Return an image with a grid overlay, using the provided spacing.

    Parameters
    ----------
    image : array, shape (M, N, 3)
        The input image.
    spacing : int
        The spacing between the grid lines.

    Returns
    -------
    image_gridded : array, shape (M, N, 3)
        The original image with a blue grid superimposed.
    """
    image_gridded = image.copy()
    image_gridded[spacing:-1:spacing, :] = [0, 0, 255]
    image_gridded[:, spacing:-1:spacing] = [0, 0, 255]
    return image_gridded

plt.imshow(overlay_grid(astro, 128));
```

Figure A-1. Astronaut image overlaid with a grid

Note that we used -1 to mean the last value of the axis, as is standard in Python indexing. You can omit this value, but the meaning is slightly different. Without it (i.e., spacing::spacing), you go all the way to the end of the array, including the final row/column. When you use it as the stop index, you prevent the final row from being selected. In the case of a grid overlay, this is probably the desired behavior.

Solution: Conway's Game of Life

This is the solution for "Exercise: Conway's Game of Life" on page 67.

Nicolas Rougier (*https://github.com/rougier*) (@NPRougier) provides a NumPy-only solution in Exercise 79 on his 100 NumPy Exercises page (*http://www.labri.fr/perso/nrougier/teaching/numpy.100/*):

```
def next_generation(Z):
    N = (Z[0:-2,0:-2] + Z[0:-2,1:-1] + Z[0:-2,2:] +
         Z[1:-1,0:-2]                + Z[1:-1,2:] +
         Z[2:  ,0:-2] + Z[2:  ,1:-1] + Z[2:  ,2:])

    # Apply rules
    birth = (N==3) & (Z[1:-1,1:-1]==0)
    survive = ((N==2) | (N==3)) & (Z[1:-1,1:-1]==1)
    Z[...] = 0
    Z[1:-1,1:-1][birth | survive] = 1
    return Z
```

Then we can start a board with:

```
random_board = np.random.randint(0, 2, size=(50, 50))
n_generations = 100
for generation in range(n_generations):
    random_board = next_generation(random_board)
```

Using a generic filter makes it even easier:

```
def nextgen_filter(values):
    center = values[len(values) // 2]
    neighbors_count = np.sum(values) - center
    if neighbors_count == 3 or (center and neighbors_count == 2):
        return 1.
    else:
        return 0.

def next_generation(board):
    return ndi.generic_filter(board, nextgen_filter,
                              size=3, mode='constant')
```

The nice thing is that some formulations of the Game of Life use what's known as a *toroidal board*, which means that the left and right ends "wrap around" and connect to each other, as well as the top and bottom ends. With `generic_filter`, it's trivial to modify our solution to incorporate this:

```
def next_generation_toroidal(board):
    return ndi.generic_filter(board, nextgen_filter,
                              size=3, mode='wrap')
```

We can now simulate this toroidal board for a few generations:

```
random_board = np.random.randint(0, 2, size=(50, 50))
n_generations = 100
for generation in range(n_generations):
    random_board = next_generation_toroidal(random_board)
```

Solution: Sobel Gradient Magnitude

This is the solution for "Exercise: Sobel Gradient Magnitude" on page 68.

```
hsobel = np.array([[ 1,  2,  1],
                   [ 0,  0,  0],
                   [-1, -2, -1]])

vsobel = hsobel.T

hsobel_r = np.ravel(hsobel)
vsobel_r = np.ravel(vsobel)

def sobel_magnitude_filter(values):
    h_edge = values @ hsobel_r
    v_edge = values @ vsobel_r
    return np.hypot(h_edge, v_edge)
```

Now we can try it out on the coins image:

```
sobel_mag = ndi.generic_filter(coins, sobel_magnitude_filter, size=3)
plt.imshow(sobel_mag, cmap='viridis');
```

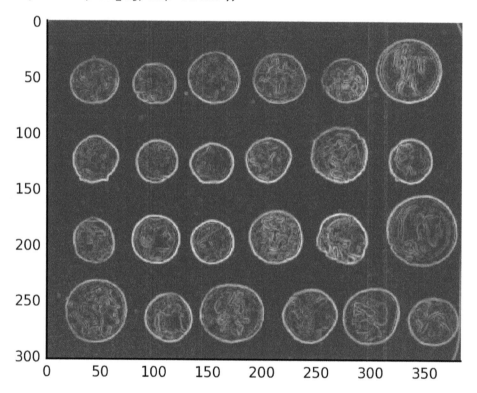

Solution: Curve Fitting with SciPy

This is the solution for "Exercise: Curve Fitting with SciPy" on page 72.

Let's look at the start of the docstring for `curve_fit`:

```
Use nonlinear least squares to fit a function, f, to data.

Assumes ``ydata = f(xdata, *params) + eps``

Parameters
----------
f : callable
    The model function, f(x, ...).  It must take the independent
    variable as the first argument and the parameters to fit as
    separate remaining arguments.
xdata : An M-length sequence or an (k,M)-shaped array
    for functions with k predictors.
    The independent variable where the data is measured.
ydata : M-length sequence
    The dependent data --- nominally f(xdata, ...)
```

It looks like we just need to provide a function that takes in a data point and some parameters, and returns the predicted value. In our case, we want the cumulative remaining frequency, $f(d)$, to be proportional to $d^{-\gamma}$. That means we need $f(d) = \alpha d^{-gamma}$:

```
def fraction_higher(degree, alpha, gamma):
    return alpha * degree ** (-gamma)
```

Then, we need our X and Y data to fit, for $d > 10$:

```
x = 1 + np.arange(len(survival))
valid = x > 10
x = x[valid]
y = survival[valid]
```

We can now use curve_fit to obtain fit parameters:

```
from scipy.optimize import curve_fit

alpha_fit, gamma_fit = curve_fit(fraction_higher, x, y)[0]
```

Let's plot the results to see how we did:

```
y_fit = fraction_higher(x, alpha_fit, gamma_fit)

fig, ax = plt.subplots()
ax.loglog(np.arange(1, len(survival) + 1), survival)
ax.set_xlabel('in-degree distribution')
ax.set_ylabel('fraction of neurons with higher in-degree distribution')
ax.scatter(avg_in_degree, 0.0022, marker='v')
ax.text(avg_in_degree - 0.5, 0.003, 'mean=%.2f' % avg_in_degree)
ax.set_ylim(0.002, 1.0)
ax.loglog(x, y_fit, c='red');
```

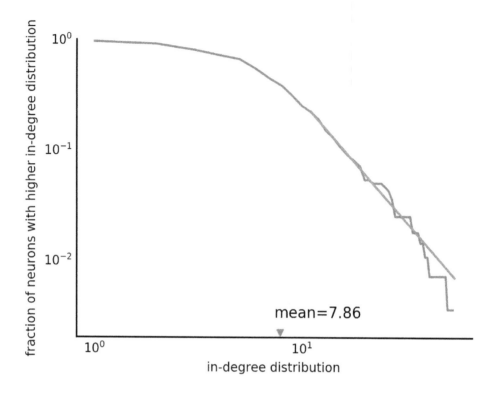

Voilà! A full Figure 6B, fit and all!

Solution: Image Convolution

This is the solution to "Exercise: Image Convolution" on page 123.

```
from scipy import signal

x = np.random.random((50, 50))
y = np.ones((5, 5))

L = x.shape[0] + y.shape[0] - 1
Px = L - x.shape[0]
Py = L - y.shape[0]

xx = np.pad(x, ((0, Px), (0, Px)), mode='constant')
yy = np.pad(y, ((0, Py), (0, Py)), mode='constant')

zz = np.fft.ifft2(np.fft.fft2(xx) * np.fft.fft2(yy)).real
print('Resulting shape:', zz.shape, ' <-- Why?')

z = signal.convolve2d(x, y)
```

```
print('Results are equal?', np.allclose(zz, z))

Resulting shape: (54, 54)  <-- Why?
Results are equal? True
```

Solution: Computational Complexity of Confusion Matrices

This is the solution for "Exercise: Computational Complexity of Confusion Matrices" on page 128.

From Chapter 1, you recall that `arr == k` creates an array of Boolean (`True` or `False`) values of the same size as `arr`. This, as you might expect, requires a full pass over `arr`. Therefore, in the above solution, we make a full pass over each of `pred` and `gt` for every combination of values in `pred` and `gt`. In principle, we can compute `cont` using just a single pass over both arrays, so these multiple passes are inefficient.

Solution: Alternative Confusion Matrix Computing

This is the solution to "Exercise: Alternative Algorithm to Compute the Confusion Matrix" on page 128.

We offer two solutions here, although many are possible.

Our first solution uses Python's built-in `zip` function to pair together labels from `pred` and `gt`.

```
def confusion_matrix1(pred, gt):
    cont = np.zeros((2, 2))
    for i, j in zip(pred, gt):
        cont[i, j] += 1
    return cont
```

Our second solution is to iterate over all possible indices of `pred` and `gt` and manually grab the corresponding value from each array:

```
def confusion_matrix1(pred, gt):
    cont = np.zeros((2, 2))
    for idx in range(len(pred)):
        i = pred[idx]
        j = gt[idx]
        cont[i, j] += 1
    return cont
```

The first option might be considered the more "Pythonic" of the two, but the second one is easier to speed up by translating and compiling in languages or tools such as C, Cython, and Numba (a topic for another book).

Solution: Computing the Confusion Matrix

This is the solution to "Exercise: Multiclass Confusion Matrix" on page 128.

We merely need to make an initial pass through both input arrays to determine the maximum label. We then add 1 to it to account for the zero label and Python's 0-based indexing. We then create the matrix and fill it in the same way as above:

```
def general_confusion_matrix(pred, gt):
    n_classes = max(np.max(pred), np.max(gt)) + 1
    cont = np.zeros((n_classes, n_classes))
    for i, j in zip(pred, gt):
        cont[i, j] += 1
    return cont
```

Solution: COO Representation

This is the solution to "Exercise: COO Representation" on page 130.

We first list the nonzero elements of the array, left to right and top to bottom, as if reading a book:

```
s2_data = np.array([6, 1, 2, 4, 5, 1, 9, 6, 7])
```

We then list the row indices of those values in the same order:

```
s2_row = np.array([0, 1, 1, 1, 1, 2, 3, 4, 4])
```

And finally the column indices:

```
s2_col = np.array([2, 0, 1, 3, 4, 1, 0, 3, 4])
```

We can easily check that these produce the right matrix, by checking equality in both directions:

```
s2_coo0 = sparse.coo_matrix(s2)
print(s2_coo0.data)
print(s2_coo0.row)
print(s2_coo0.col)

[6 1 2 4 5 1 9 6 7]
[0 1 1 1 1 2 3 4 4]
[2 0 1 3 4 1 0 3 4]
```

and:

```
s2_coo1 = sparse.coo_matrix((s2_data, (s2_row, s2_col)))
print(s2_coo1.toarray())

[[0 0 6 0 0]
 [1 2 0 4 5]
 [0 1 0 0 0]
 [9 0 0 0 0]
 [0 0 0 6 7]]
```

Solution: Image Rotation

This is the solution to "Exercise: Image Rotation" on page 138.

We can *compose* transformations by multiplying them. We know how to rotate an image about the origin, as well as how to slide it around. So what we will do is slide the image so that the center is at the origin, rotate it, and then slide it back.

```
def transform_rotate_about_center(shape, degrees):
    """Return the homography matrix for a rotation about an image center."""
    c = np.cos(np.deg2rad(angle))
    s = np.sin(np.deg2rad(angle))

    H_rot = np.array([[c, -s,  0],
                      [s,  c,  0],
                      [0,  0,  1]])
    # compute image center coordinates
    center = np.array(image.shape) / 2
    # matrix to center image on origin
    H_tr0 = np.array([[1, 0, -center[0]],
                      [0, 1, -center[1]],
                      [0, 0,          1]])
    # matrix to move center back
    H_tr1 = np.array([[1, 0, center[0]],
                      [0, 1, center[1]],
                      [0, 0,         1]])
    # complete transformation matrix
    H_rot_cent = H_tr1 @ H_rot @ H_tr0

    sparse_op = homography(H_rot_cent, image.shape)

    return sparse_op
```

We can test that this works:

```
tf = transform_rotate_about_center(image.shape, 30)
plt.imshow(apply_transform(image, tf));
```

Solution: Reducing the Memory Footprint

This is the solution to "Exercise: Reducing the Memory Footprint" on page 140.

The np.ones array that we create is read-only: it will only be used as the values to sum by coo_matrix. We can use broadcast_to to create a similar array with only one element, "virtually" repeated n times:

```
def confusion_matrix(pred, gt):
    n = pred.size
    ones = np.broadcast_to(1., n)  # virtual array of 1s of size n
    cont = sparse.coo_matrix((ones, (pred, gt)))
    return cont
```

Let's make sure it still works as expected:

```
cont = confusion_matrix(pred, gt)
print(cont.toarray())

[[ 3.  1.]
 [ 2.  4.]]
```

Boom. Instead of making an array as big as the original data, we just make one of size 1. As we handle bigger and bigger datasets, such optimizations become increasingly important.

Solution: Computing Conditional Entropy

To obtain the joint probability table, we simply divide the table by its total—in this case, 12:

This is the solution to "Exercise: Computing Conditional Entropy" on page 144.

```
print('table total:', np.sum(p_rain_g_month))
p_rain_month = p_rain_g_month / np.sum(p_rain_g_month)

table total: 12.0
```

Now we can compute the conditional entropy of the month given rain. (This is like asking: if we know it's raining, how much more information do we need to know to figure out what month it is, on average?)

```
p_rain = np.sum(p_rain_month, axis=0)
p_month_g_rain = p_rain_month / p_rain
Hmr = np.sum(p_rain * p_month_g_rain * np.log2(1 / p_month_g_rain))
print(Hmr)

3.5613602411
```

Let's compare that to the entropy of the months:

```
p_month = np.sum(p_rain_month, axis=1)  # 1/12, but this method is more general
Hm = np.sum(p_month * np.log2(1 / p_month))
print(Hm)

3.58496250072
```

So we can see that knowing whether it rained today got us two hundredths of a bit closer to guessing what month it is! Don't bet the farm on that guess.

Solution: Rotation Matrix

This is the solution to "Exercise: Rotation Matrix" on page 159.

Part 1:

```
import numpy as np

theta = np.deg2rad(45)
```

```
R = np.array([[np.cos(theta), -np.sin(theta), 0],
              [np.sin(theta),  np.cos(theta), 0],
              [      0,              0, 1]])

print("R times the x-axis:", R @ [1, 0, 0])
print("R times the y-axis:", R @ [0, 1, 0])
print("R times a 45 degree vector:", R @ [1, 1, 0])

R times the x-axis: [ 0.70710678  0.70710678  0.        ]
R times the y-axis: [-0.70710678  0.70710678  0.        ]
R times a 45 degree vector: [  1.11022302e-16   1.41421356e+00   0.00000000e+00]
```

Part 2:

Since multiplying a vector by R rotates it 45 degrees, multiplying the result by R again should result in the original vector being rotated 90 degrees. Matrix multiplication is associative, which means that $R(Rv) = (RR)v$, so $S = RR$ should rotate vectors by 90 degrees around the z-axis.

```
S = R @ R
S @ [1, 0, 0]

array([  2.22044605e-16,   1.00000000e+00,   0.00000000e+00])
```

Part 3:

```
print("R @ z-axis:", R @ [0, 0, 1])

R @ z-axis: [ 0.  0.  1.]
```

R rotates both the x- and y-axes, but not the z-axis.

Part 4:

Looking at the documentation of eig, we see that it returns two values: a 1D array of eigenvalues, and a 2D array where each column contains the eigenvector corresponding to each eigenvalue.

```
np.linalg.eig(R)

(array([ 0.70710678+0.70710678j,  0.70710678-0.70710678j,  1.00000000+0.j      ]),
 array([[ 0.70710678+0.j      ,  0.70710678-0.j      ,  0.00000000+0.j      ],
        [ 0.00000000-0.70710678j,  0.00000000+0.70710678j,  0.00000000+0.j      ],
        [ 0.00000000-0.j      ,  0.00000000+0.j      ,  1.00000000+0.j      ]]))
```

In addition to some complex-valued eigenvalues and vectors, we see the value 1 associated with the vector $[0, 0, 1]^T$.

Solution: Showing the Affinity View

This is the solution for "Exercise: Showing the Affinity View" on page 170.

In the affinity view, instead of using the processing depth on the y-axis, we use the normalized third eigenvector of Q, just like we did with x. (And we invert it if necessary, just like we did with x!)

```python
y = Dinv2 @ Vec[:, 2]
asjl_index = np.argwhere(neuron_ids == 'ASJL')
if y[asjl_index] < 0:
    y = -y
```

```python
plot_connectome(x, y, C, labels=neuron_ids, types=neuron_types,
                type_names=['sensory neurons', 'interneurons',
                            'motor neurons'],
                xlabel='Affinity eigenvector 1',
                ylabel='Affinity eigenvector 2')
```

Challenge Accepted: Linear Algebra with Sparse Matrices

This is the solution for "Exercise Challenge: Linear Algebra with Sparse Matrices" on page 170.

For the purposes of this challenge, we are going to use the small connectome, because it's easier to visualize what is going on. In later parts of the challenge we'll use these techniques to analyze larger networks.

First, we start with the adjacency matrix, A, in a sparse matrix format, in this case CSR, which is the most common format for linear algebra. We'll append s to the names of all the matrices to indicate that they are sparse.

```
from scipy import sparse
import scipy.sparse.linalg

As = sparse.csr_matrix(A)
```

We can create our connectivity matrix in the same way:

```
Cs = (As + As.T) / 2
```

In order to get the degrees matrix, we can use the "diags" sparse format, which stores diagonal and off-diagonal matrices.

```
degrees = np.ravel(Cs.sum(axis=0))
Ds = sparse.diags(degrees)
```

Getting the Laplacian is straightforward:

```
Ls = Ds - Cs
```

Now we want to get the processing depth. Remember that getting the pseudoinverse of the Laplacian matrix is out of the question, because it will be a dense matrix (the inverse of a sparse matrix is not generally sparse itself). However, we were actually using the pseudo-inverse to compute a vector z that would satisfy $Lz = b$, where $b = C \odot \text{sign}(A - A^T)\,\mathbf{1}$. (You can see this in the supplementary material for Varshney et al.) With dense matrices, we can simply use $z = L^+b$. With sparse ones, though, we can use one of the *solvers* (see "Solvers" on page 171) in sparse.linalg.isolve to get the z vector after providing L and b—no inversion required!

```
b = Cs.multiply((As - As.T).sign()).sum(axis=1)
z, error = sparse.linalg.isolve.cg(Ls, b, maxiter=10000)
```

Finally, we must find the eigenvectors of Q, the degree-normalized Laplacian, corresponding to its second and third smallest eigenvalues.

You might recall from Chapter 5 that the numerical data in sparse matrices is in the .data attribute. We use that to invert the degrees matrix:

```
Dsinv2 = Ds.copy()
Dsinv2.data = 1 / np.sqrt(Ds.data)
```

Finally, we use SciPy's sparse linear algebra functions to find the desired eigenvectors. The Q matrix is symmetric, so we can use the eigsh function, specialized for symmetric matrices, to compute them. We use the which keyword argument to specify that we want the eigenvectors corresponding to the smallest eigenvalues, and k to specify that we need the three smallest:

```
Qs = Dsinv2 @ Ls @ Dsinv2
vals, Vecs = sparse.linalg.eigsh(Qs, k=3, which='SM')
```

```
sorted_indices = np.argsort(vals)
Vecs = Vecs[:, sorted_indices]
```

Finally, we normalize the eigenvectors to get the x and y coordinates (and flip these if necessary):

```
_dsinv, x, y = (Dsinv2 @ Vecs).T
if x[vc2_index] < 0:
    x = -x
if y[asjl_index] < 0:
    y = -y
```

(Note that the eigenvector corresponding to the smallest eigenvalue is always a vector of all ones, which we're not interested in.) We can now reproduce the following plots!

```
plot_connectome(x, z, C, labels=neuron_ids, types=neuron_types,
                type_names=['sensory neurons', 'interneurons',
                            'motor neurons'],
                xlabel='Affinity eigenvector 1', ylabel='Processing depth')

plot_connectome(x, y, C, labels=neuron_ids, types=neuron_types,
                type_names=['sensory neurons', 'interneurons',
                            'motor neurons'],
                xlabel='Affinity eigenvector 1',
                ylabel='Affinity eigenvector 2')
```

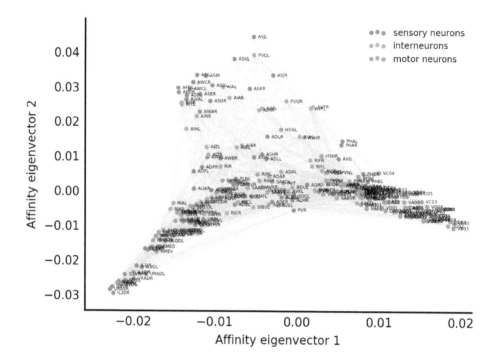

Solution: Dealing with Dangling Nodes

This is the solution for "Exercise: Dealing with Dangling Nodes" on page 176.

In order to have a stochastic matrix, all columns of the transition matrix must sum to 1. This is not satisfied when a species isn't eaten by any others: that column will consist of all zeros. Replacing all those columns by $1/n1$, however, would be expensive.

The key is to realize that *every row* will contribute the *same amount* to the multiplication of the transition matrix by the current probability vector. That is to say, adding these columns will add a single value to the result of the iteration multiplication. What value? $1/n$ times the elements of r that correspond to a dangling node. This can be expressed as a dot-product of a vector containing $1/n$ for positions corresponding to dangling nodes, and zero elsewhere, with the vector r for the current iteration.

```
def power2(Trans, damping=0.85, max_iter=10**5):
    n = Trans.shape[0]
    dangling = (1/n) * np.ravel(Trans.sum(axis=0) == 0)
    r0 = np.full(n, 1/n)
    r = r0
    for _ in range(max_iter):
        rnext = (damping * (Trans @ r + dangling @ r) +
                (1 - damping) / n)
        if np.allclose(rnext, r):
```

```
            return rnext
    else:
        r = rnext
return r
```

Try this out manually for a few iterations. Notice that if you start with a stochastic vector (a vector whose elements all sum to 1), the next vector will still be a stochastic vector. Thus, the output PageRank from this function will be a true probability vector, and the values will represent the probability that we end up at a particular species when following links in the food chain.

Solution: Verify Methods

This is the solution for "Exercise: Equivalence of Different Eigenvector Methods" on page 176.

np.corrcoef gives the Pearson correlation coefficient between all pairs of a list of vectors. This coefficient will be equal to 1 if and only if two vectors are scalar multiples of each other. Therefore, a correlation coefficient of 1 is sufficient to show that the above methods produce the same ranking.

```
pagerank_power = power(Trans)
pagerank_power2 = power2(Trans)
np.corrcoef([pagerank, pagerank_power, pagerank_power2])

array([[ 1.,   1.,   1.],
       [ 1.,   1.,   1.],
       [ 1.,   1.,   1.]])
```

Solution: Modify the align Function

This is the solution for "Exercise: Modify the align Function" on page 190.

We use basin hopping at the higher levels of the pyramid, but use Powell's method for the lower levels, because basin hopping is too computationally expensive to run at full resolution:

```
def align(reference, target, cost=cost_mse, nlevels=7, method='Powell'):
    pyramid_ref = gaussian_pyramid(reference, levels=nlevels)
    pyramid_tgt = gaussian_pyramid(target, levels=nlevels)

    levels = range(nlevels, 0, -1)
    image_pairs = zip(pyramid_ref, pyramid_tgt)

    p = np.zeros(3)

    for n, (ref, tgt) in zip(levels, image_pairs):
        p[1:] *= 2
        if method.upper() == 'BH':
            res = optimize.basinhopping(cost, p,
```

```
                                    minimizer_kwargs={'args': (ref, tgt)})
            if n <= 4:  # avoid basin hopping in lower levels
                method = 'Powell'
        else:
            res = optimize.minimize(cost, p, args=(ref, tgt), method='Powell')
        p = res.x
        # print current level, overwriting each time (like a progress bar)
        print(f'Level: {n}, Angle: {np.rad2deg(res.x[0]) :.3}, '
              f'Offset: ({res.x[1] * 2**n :.3}, {res.x[2] * 2**n :.3}), '
              f'Cost: {res.fun :.3}', end='\r')

    print('')  # newline when alignment complete
    return make_rigid_transform(p)
```

Now let's try that alignment:

```
from skimage import util

theta = 50
rotated = transform.rotate(astronaut, theta)
rotated = util.random_noise(rotated, mode='gaussian',
                            seed=0, mean=0, var=1e-3)

tf = align(astronaut, rotated, nlevels=8, method='BH')
corrected = transform.warp(rotated, tf, order=3)

f, (ax0, ax1, ax2) = plt.subplots(1, 3)
ax0.imshow(astronaut)
ax0.set_title('Original')
ax1.imshow(rotated)
ax1.set_title('Rotated')
ax2.imshow(corrected)
ax2.set_title('Registered')
for ax in (ax0, ax1, ax2):
    ax.axis('off')
```

```
Level: 1, Angle: -50.0, Offset: (-2.09e+02, 5.74e+02), Cost: 0.0385
```

Original

Rotated

Registered

Success! Basin hopping was able to recover the correct alignment, even in the problematic case in which the `minimize` function got stuck.

Solution: scikit-learn Library

This is the solution to "Exercise: PCA of Streaming Data" on page 214.

First, we write the function to train the model. The function should take in a stream of samples and output a PCA model, which can *transform* new samples by projecting them from the original n-dimensional space to the principal component space.

```python
import toolz as tz
from toolz import curried as c
from sklearn import decomposition
from sklearn import datasets
import numpy as np

def streaming_pca(samples, n_components=2, batch_size=100):
    ipca = decomposition.IncrementalPCA(n_components=n_components,
                                        batch_size=batch_size)
    tz.pipe(samples,  # iterator of 1D arrays
            c.partition(batch_size),  # iterator of tuples
            c.map(np.array),  # iterator of 2D arrays
            c.map(ipca.partial_fit),  # partial_fit on each
            tz.last)  # Suck the stream of data through the pipeline
    return ipca
```

Now, we can use this function to *train* (or *fit*) a PCA model:

```python
reshape = tz.curry(np.reshape)

def array_from_txt(line, sep=',', dtype=np.float):
    return np.array(line.rstrip().split(sep), dtype=dtype)

with open('data/iris.csv') as fin:
    pca_obj = tz.pipe(fin, c.map(array_from_txt), streaming_pca)
```

Finally, we can stream our original samples through the `transform` function of our model. We stack them together to obtain a `n_samples` by `n_components` matrix of data:

```python
with open('data/iris.csv') as fin:
    components = tz.pipe(fin,
                         c.map(array_from_txt),
                         c.map(reshape(newshape=(1, -1))),
                         c.map(pca_obj.transform),
                         np.vstack)

print(components.shape)
(150, 2)
```

We can now plot the components:

```
iris_types = np.loadtxt('data/iris-target.csv')
plt.scatter(*components.T, c=iris_types, cmap='viridis');
```

You can verify that this gives (approximately) the same result as a standard PCA (compare Figures A-2 and A-3):

```
iris = np.loadtxt('data/iris.csv', delimiter=',')
components2 = decomposition.PCA(n_components=2).fit_transform(iris)
plt.scatter(*components2.T, c=iris_types, cmap='viridis');
```

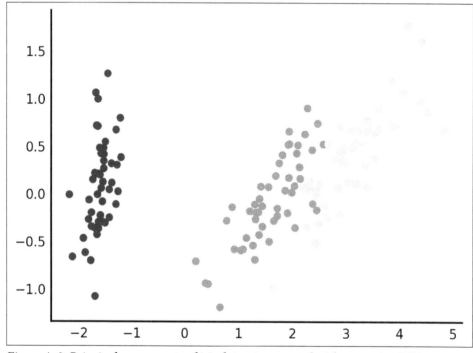

Figure A-2. Principal components of iris dataset computed with streaming PCA

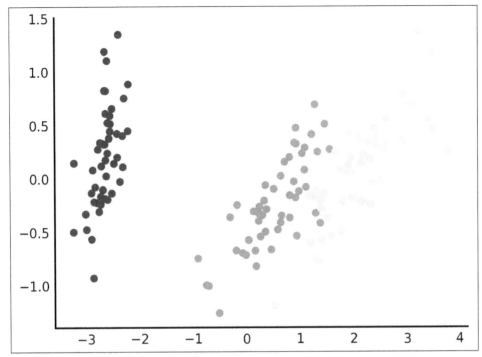

Figure A-3. Principal components of iris dataset computed with normal PCA

The difference, of course, is that streaming PCA can scale to extremely large datasets.

Solution: Add a Step to the Start of the Pipe

This is the solution to "Exercise: Online Unzip" on page 217.

We can replace open in the original genome code with a curried version of gzip.open. The default mode of gzip's open function is rb (read **b**ytes), instead of rt for Python's built-in open (read **t**ext), so we have to provide it.

```
import gzip

gzopen = tz.curry(gzip.open)


def genome_gz(file_pattern):
    """Stream a genome, letter by letter, from a list of FASTA filenames."""
    return tz.pipe(file_pattern, glob, sorted,  # Filenames
                c.map(gzopen(mode='rt')),  # lines
                # concatenate lines from all files:
                tz.concat,
                # drop header from each sequence
                c.filter(is_sequence),
```

```
# concatenate characters from all lines
tz.concat,
# discard newlines and 'N'
c.filter(is_nucleotide))
```

You can try this out with the compressed drosophila genome file:

```
dm = 'data/dm6.fa.gz'
model = tz.pipe(dm, genome_gz, c.take(10**7), markov)
plot_model(model, labels='ACGTacgt')
```

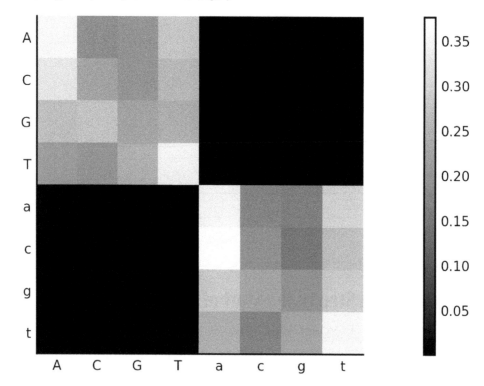

If you want to have a single genome function, you could write a custom open function that decides from the filename, or from trying and failing, whether a file is a *gzip* file.

Similarly, if you have a *.tar.gz* full of FASTA files, you can use Python's tarfile module instead of glob to read each file individually. The only caveat is that you will have to use the bytes.decode function to decode each line, as tarfile returns them as bytes, not as text.

Index

DNA (deoxyribonucleic acid), 3

About the Authors

Juan Nunez-Iglesias is a freelance consultant and a Research Scientist at the University of Melbourne, Australia. Prior positions include Research Associate at HHMI Janelia Farm (where he worked with Mitya Chklovskii) and Research Assistant/PhD student at the University of Southern California (where he studied computational biology supervised by Xianghong Jasmine Zhou). His principal research interests are neuroscience and image analysis. He is also interested in graph methods in bioinformatics and in biostatistics.

Stéfan van der Walt is an assistant researcher at the Berkeley Institute for Data Science at the University of California, Berkeley, and a senior lecturer in applied mathematics at Stellenbosch University, South Africa. He has been involved in the development of scientific open source software for more than a decade, and enjoys teaching Python at workshops and conferences. Stéfan is the founder of scikit-image and a contributor to NumPy, SciPy, and cesium-ml.

Harriet Dashnow is a bioinformatician and has worked at the Murdoch Childrens Research Institute, the Department of Biochemistry at the University of Melbourne, and the Victorian Life Sciences Computation Initiative (VLSCI). Harriet obtained a BA (Psychology), a BS (Genetics and Biochemistry), and a MS (Bioinformatics) from the University of Melbourne. She is currently working toward a PhD. She organizes and teaches computational skills workshops in such areas as genomics, Software Carpentry, Python, R, Unix, and Git version control.

Colophon

The animal on the cover of *Elegant SciPy* is a paradise whydah (*Vidua paradisaea*), or the long-tailed paradise whydah. This small sparrow-like bird is found in Eastern Africa through South Sudan and southern Angola.

Male and female paradise whydahs are almost indistinguishable until breeding season when males moult into their breeding plumage. Male breeding plumage includes a black head, brown breast, bright yellow plumage around the nape of the neck, and a white abdomen and long, broad black tail feathers approximately three times the length of its body.

Paradise whydahs are brood parasites to green-winged pytilia. Males will imitate the song of the male pytilia and because they are louder and larger, the green-winged foster parents will give them more attention. This brood parasitic nature makes them difficult to breed in captivity, however, males of this species are often sold as pets in the U.S. and other countries. The long-tailed paradise whydah is evaluated as Least Concern.

Many of the animals on O'Reilly covers are endangered; all of them are important to the world. To learn more about how you can help, go to *animals.oreilly.com*.

The cover image is from *Wood's Illustrated Natural History*. The cover fonts are URW Typewriter and Guardian Sans. The text font is Adobe Minion Pro; the heading font is Adobe Myriad Condensed; and the code font is Dalton Maag's Ubuntu Mono.

Learn from experts.
Find the answers you need.

Sign up for a **10-day free trial** to get **unlimited access** to all of the content on Safari, including Learning Paths, interactive tutorials, and curated playlists that draw from thousands of ebooks and training videos on a wide range of topics, including data, design, DevOps, management, business—and much more.

Start your free trial at:

oreilly.com/safari

(No credit card required.)